AF333980

BIOINFORMATICS

Edited by **Horacio Pérez-Sánchez**

Bioinformatics
http://dx.doi.org/10.5772/3089
Edited by Horacio Pérez-Sánchez

Contributors

Shubhalaxmi Kher, Jianling Peng, Eve Syrkin Wurtele, Julie Dickerson, Mohd Fakharul Zaman Raja Yahya, Umi Marshida Abdul Hamid, Farida Zuraina Mohd Yusof, Felipe García-Vallejo, Martha Cecilia Domínguez, Matthew Ezewudo, Promita Bose, Kajari Mondal, Viren Patel, Dhanya Ramachandran, Michael E. Zwick, Hugo Saldanha, Edward Ribeiro, Carlos Borges, Aletéia Araújo, Ricardo Gallon, Maristela Holanda, Maria Emília Walter, Roberto Togawa, João Carlos Setubal, Imre Pechan, Béla Fehér, Ly Le, María J. R. Yunta, León P. Martínez-Castilla, Rogelio Rodríguez-Sotres, Scheila de Avila e Silva, Sergio Echeverrigaray, Suman Ghosal, Shaoli Das, Jayprokas Chakrabarti, Yoshiaki Mizuguchi, Takuya Mishima, Eiji Uchida, Toshihiro Takizawa, Harun Pirim, Şadi Evren Şeker, Haiping Wang, Taining Xiang, Xuegang Hua

Published by InTech

Janeza Trdine 9, 51000 Rijeka, Croatia

Edition 2014

Copyright © 2012 InTech

Notice

Publishing Process Manager Marina Jozipovic
Technical Editor InTech DTP team
Cover InTech Design team

Additional hard copies can be obtained from orders@intechopen.com

Bioinformatics, Edited by Horacio Pérez-Sánchez
p. cm.
ISBN 978-953-51-0878-8

Contents

Analysis of Biological Networks

Chapter 1

Hierarchical Biological Pathway Data Integration and Mining

Shubhalaxmi Kher, Jianling Peng, Eve Syrkin Wurtele and Julie Dickerson

Additional information is available at the end of the chapter

http://dx.doi.org/10.5772/49974

1. Introduction

Biological pathway data is the key resource for biologists worldwide. Interestingly, most of these sources that generate, update, and analyze data are open source. One of the observations that motivated this research work is that, the repositories of data created by a variety of laboratories and research units worldwide represent same pathways with significant details. Generally, if the pathway data has resulted from experimentation, then it is expected that across different resources, under similar conditions, pathways would be exactly identical and biologists may pickup from any source. Interestingly, almost all of the biological data sources refer to data integration of some kind. It may involve rigorous integration mechanisms within the data source and the purpose of integration may change the perspective of looking at the integration.

These efforts in integration may be either local to the source or lack details associated with integration within a pathway, across pathways, or from various data sources etc. Further, the key attributes or design criteria may not be well documented and or may not be readily available to the biologist. In other words, the integration may be achieved as vertical integration (within the data source), or horizontal integration (across data sources). Since most of the extensively integrated data sources (plants or humans) like BioCyc-level-I, Reactome are human curated, it is hard to identify the integration done by the sources like; BioCyc. Also, on a similar note, it may not be apparent to find exactly when the data was integrated looking at a pathway.

Data in general refers to a collection of results, including the results of experience, observation, or experiment, or a set of premises and can be utilized at the maximum when made available to all in a common format. Different organizations and research laboratories around the world store the data in their own formats; this diversity of data sources is caused due to many factors including lack of coordination among the organizations and research

---

laboratories. These intellectual gaps can be bridged by adopting new technology, mergers, acquisitions, and geographic coordination of collaborating groups [1].

For the open source biological databases, it is common for the biologists and researchers to refer to many databases in order to pursue inference or analysis; though it is one of the most challenging tasks. Biological pathway data integration is aimed to work with repositories of data from a variety of sources. As such, two or more databases may not provide identical information for a given pathway, but integrating these two databases may yield a richer resource for analysis. Additionally, the conditions under which data is collected, either by experimentation or by collecting evidence of the published material, in either case the supporting references play a crucial role and is of interest to the biologists in making the analysis more meaningful. At present there are over 200 biological pathway databases. However, very few of them are independently created. Some of these databases may be derived from different data sources. Unfortunately, the documentation often does not reveal details of the data collection, sources, and dates. Further, the research groups involved in analysis of the data usually selectively use data from a single data source. For example, for yeast studies, the Saccharomyces Genome Database (SGD) is the reference for most analyses [2].

In case of biological pathway data, rapid accumulation of genomic and proteomic data have made two major bioinformatics problems apparent.

- The lack of communication between different bioinformatics data resources; whether they are databases or individual analysis programs.
- Biological data are hierarchical and highly related yet are conventionally stored separately in individual database and in different formats.
- Additionally, they are governed more by how data is obtained rather than by what they mean.

Most commercially available bioinformatics systems perform functional analysis using a single data source; an approach that emphasizes pathway mapping and relationship inference based on the data acquired from multiple data sources. Each pathway modality in the data has its own specific representation issues which must be understood before attempting to integrate across modalities.

1.1. Overview

There has been a dramatic increase in the number of large scale comprehensive biological databases that provide useful resources to the community like; Biochemical Pathways (KEGG, AraCyc, and MapMan), Protein Interactions (biomolecular interaction network database), or systems like; Dragon Plant Biology Explorer and Pathway Miner for integrating associations in metabolic networks and ontologies [3-8]. Other databases such as Regulon DB, PlantCARE, PLACE, EDP:Eurokaryotic promoter database, Transcription Regulatory Regions Database, Athamap, and TRANSFAC store information related to transcriptional regulation[9-15].

The aim of molecular biology is to understand the regulation of protein synthesis and its reactions to external and internal signals. All the cells in an organism carry the same genomic data, yet their protein makeup can be drastically different; both temporally and spatially, due to regulation. Protein synthesis is regulated by many mechanisms at its different stages. These include mechanisms for controlling transcription initiation, RNA splicing, mRNA transport, translation initiation, post-translational modifications, and degradation of mRNA/protein. One of the main junctions at which regulation occurs is mRNA transcription. A major role in this machinery is played by proteins themselves that bind to regulatory regions along the DNA, greatly affecting the transcription of the genes they regulate [16]. Friedman introduces a new approach for analyzing gene expression patterns that uncovers properties of the transcriptional program by examining statistical properties of dependence and conditional independence in the data.

For protein interactions, it is intended to connect related proteins and link biological functions in the context of larger cellular processes [17]. The content of these data sources typically complements the experimentally determined protein interactions with the ones that are predicted from gene proximity, fusion, co-expressed data, as well as those determined by using phylogenetic profiling. Each pathway modality in the data has its own specific representation issues which must be understood before integration across modalities is attempted. At present, the bioinformatics database owner only develops private system to provide user with data query and analysis services; such as NCBI develops Entrez database query system which is used on GenBank. European Molecular Biology Laboratory (EMBL) develops Sequence Retrieval Systems. The EMBL Nucleotide Sequence Database maintained at the European Bioinformatics Institute (EBI), incorporates, organizes, and distributes nucleotide sequences from public sources [18]. The database is a part of an international collaboration with DDBJ (Japan) and GenBank (USA). Data are exchanged between the collaborating databases on a daily basis to achieve optimal synchrony. The key point is how to share the heterogeneous databases and make a common query platform for users [19].

Friedman [16] describes early microarray experiments that examined few samples and mainly focused on differential display across tissues or conditions of interest. Such experiments collect enormous amounts of data, which clearly reflects many aspects of the underlying biological processes. An important challenge is to develop methodologies that are both statistically sound and computationally tractable for analyzing such data sets and inferring biological interactions from them. Most of the analysis tools currently used are based on clustering algorithms. The clustering algorithms attempt to locate groups of genes that have similar expression patterns over a set of experiments. Such analysis has proven to be useful in discovering genes that are co-regulated and/or have similar function. A more ambitious goal for analysis is to reveal the structure of the transcriptional regulation process. This is clearly a hard problem. Not only the current data is extremely noisy, but, mRNA expression data alone only gives a partial picture that does not reflect key events such as; translation and protein (in) activation. Finally, the amount of samples, even in the largest experiments in the foreseeable future, does not provide enough information to construct a fully detailed model with high statistical significance.

sources and then displays the fetched data for its user base. Queries in federated databases are executed within remote data sources and results displayed in federated databases are extracted remotely from the data sources. Due to this capability, federated databasing has two major advantages.

- Federated databases can be regarded as an on-demand approach to provide immediate access to up-to-date data deposited in multiple data sources.
- Compared with data warehousing, federated databasing does not replicate data in data sources; therefore, it presents relatively inexpensive costs for storage and curation. However, federated databasing still has to update its query translation to keep pace with data access methods at diverse remote data sources.

Service –Oriented Approach

A decentralized approach is also being developed, in which individual data sources agree to open their data via Web Services (WS). The service-oriented approach enables data integration from multiple heterogeneous data sources through computer interoperability. The service-oriented approach features data integration through computer-to-computer communication via Web API and up-to-date data retrieval from diverse data sources. Heterogeneous data integration requires that many data sources should become service providers by opening their data via WS and by standardizing data identities and nomenclature to ease data exchange and analysis.

Semantic Web

Most web pages in biological data sources are designed for human reading. RDF provides standard formats for data interchange and describes data as a simple statement, containing a set of triples: a subject, a predicate, and an object. Any two statements can be linked by an identical subject or object. OWL builds on RDF and Uniform Resource Identifier (URI) and describes data structure and meaning based on ontology, which enables automated data reasoning and inferences by computers. Application of semantic Web technologies is a significant advancement for bioinformatics, enabling automated data processing and reasoning. The semantic integration uses ontologies for data description and thus represents ontology-based integration. [27] reviews the current development of semantic network technologies and their applications to the integration of genomic and proteomic data. His work elaborates on applying a semantic network approach to modeling complex cell signaling pathways and simulating the cause-effect of molecular interactions in human macrophages. [31] Illustrates his approach by comparing federated approach versus warehousing versus semantic web using multiple sources.

Wiki-based Integration

A weakness common to all the above approaches is that the quantity of users' participations in the process is inadequate. With the increasing volume of biological data, data integration inevitably will require a large number of users' participations. A successful example that harnesses collective intelligence for data aggregation and knowledge collection is Wikipedia: an online encyclopedia that allows any user to create and edit content. It is infeasible to integrate such large amounts of data into a single point (such as a data warehouse). Data sources are developed for different purposes and fulfill different functions. Therefore, it is promising to establish an efficient way for data exchange among these distributed and heterogeneous data sources. However, a dozen of data sources are designed merely for data storage, but not for data exchange.

2.1. Survey of Pathway Databases and Integration Efforts

Table 1 below shows various data integration efforts and projects for biological pathways worldwide.

Biochemical pathways	Description
BRITE	Bio molecular Relations in Information Transmission and Expression
EcoCyc/MetaCyc	Encyclopaedia of E. coli genes and metabolism; Metabolic encyclopedia
EMP	Metabolic pathways
KEGG	Kyoto encyclopaedia of genes and genomes
Biochemical Pathways	Enzyme database and link to biochemical pathway map
Interactive Fly	Biochemical pathways in Drosophila
Metabolic Pathway	Metabolic pathways of biochemistry
Molecular interaction	Kohn molecular interaction maps
Malaria parasite	Malaria Parasite metabolic pathways
aMAZE	Protein function and biochemical pathways project at EBI
PathDB	Metabolic pathway information
UM-BBD	Microbial bio catalytic reactions and biodegradation pathways primarily for xenobiotic, chemical compounds
WIT	Function assignments to genes and the development of metabolic models
THCME Medical Biochemistry	Description of several metabolic and biochemical pathways
Signaling pathways	
Apoptosis	Pathways of apoptosis at KEGG
BBID	Database of images of biological pathways, macromolecular structures, gene families, and cellular relationships
BioCarta	Several signalling pathways
BIND	The bio molecular interaction network database

CSNDB	Cell signalling networks database
GeneNet	Information on gene networks, groups of co-ordinately working genes
GeNet	Information on functional organization of regulatory gene networks
SPAD	Signalling pathway database
STKE	Pathway information
TransPath	Pathways involved in the regulation of transcription factors
Protein-protein interactions	
Blue Print	Biological interaction database
CYGD	Protein-protein interaction map at Comprehensive Yeast Genome Database
CytoScape	Visualization and analysis of biological network
DIP	Database of interacting proteins
GenMAPP	Gene Map Annotator and Pathway Profiler
GRID	The General Repository for Interaction Datasets
Proteome Bio knowledge	Biological information about proteins comprise Incyte's Proteome Bio Knowledge Library
Protein Interaction Domains	Signal transduction
Reactome	A knowledgebase of biological processes
Yeast Interaction Pathway	PathCalling Yeast Interaction Database at Curagen

Table 1. Various Data integration Efforts

Other efforts towards designing new applications for data mining and integration at the K.U.Leuven Center for Computational Systems Biology include;

- aBandApart (2007): A software to mine MEDLINE abstracts to annotate human genome at the level of cytogenic bands.
- ReModiscovery (2006): An intuitive algorithm to correlate regulatory programs with regulators and corresponding motifs to a set of co-expressed genes
- LOOP (2007): A toll to analyze ArrayCGH loop designs. ArrayCGH is a microarray technology that can be used to detect aberrations in the ploidy of DNA segments in the genome of patients with congenital anomalies.
- SynTReN (2006): A generator of synthetic gene expression data for design and analysis of structure learning algorithms.
- BlockAligner (2005): Provides an API in R to query BioMart databases such as Ensemble.
- BlockSampler (2005): Finds conserved blocks in the upstream region of sets of orthologous genes.

- M@cBETH (2005) (a Microarray Classification Benchmarking Tool on a host server): Web service offers the microarray community a simple tool for making optimal two class predictions.
- TxTGate (2004): A literature index database designed towards the summarization and analysis of groups of genes based on text.
- Endeavour is a software application for the computational prioritization of test genes based on training genes using different information sources such as MEDLINE abstracts and LocusLink textual description, gene ontology, annotation, BIND protein interactions, and Transcription Factor Binding Sites (TFBS).
- TOUCAN2 (2004): A workbench for regulatory sequence analysis on metazoan genomes: Comparative genomics detection of significant transcription factor binding sites and detection of cis-regulatory modules in sets of coexpressed/ coregulated genes.
- INCLUSive (2003): A suit of algorithms and tools for the analysis of gene expression data and the directory of cis-regulatory sequence elements.
- Adaptive Quality-Based Clustering (AQBC) (2002): AQBC is a heuristic, iterative two-step algorithm to cluster gene expression data.
- MotifSampler (2001): Finds over represented motifs in the upstream region of a set of co-regulated genes.

2.2. Types of pathways

Biological networks are studied and modeled at different description levels establishing different pathway types, For example; metabolic pathways describe the conversion of metabolites by enzyme-catalyzed chemical reactions given by their stoichiometric equations, such as the main pathways of the energy household as Glycolysis or Pentose Phosphate pathway. Another pathway type is signal transduction pathways, also known as information metabolism, explaining how cells receive, process, and responds to information from the environment. A brief description about various types of pathways is given below.

A. Metabolic Pathways describe the network of enzyme-catalyzed reactions that release energy by breaking down nutrients (catabolism) and building up the essential compounds necessary for growth (anabolism). Experimentally determined metabolic pathways have established for a few model organisms, but most metabolic pathways databases contain pathway data that has been computationally inferred from the genomes annotations. Because most genome annotations are incomplete, metabolic pathway databases contain pathway holes which can only be addressed by experiment or computational inference. A good test of a reconstructed metabolic network is to ask if it can produce the set of essential compounds necessary for growth, given a known minimal nutrient set. To solve this problem, metabolism can be represented as a bipartite directed graph, where one set of nodes represents metabolites, the other set represents biochemical reactions with labeled edges used to indicate relationships between nodes (reaction X produces metabolite Y, or metabolite Y is-consumed-by reaction X.

Assign: **Direct** = *1.0*; **Indirect** = *0.8*; **Computational** =*0.6*; **Hypothetical** = *0.5*.

Step 2. EV Unification (Rule Set –I)

BioCyc is a collection of 371 pathway/genome databases. Each pathway/genome database in the BioCyc collection describes the genome and metabolic pathways of a single organism. It considers a class hierarchy with four main classes. Since BioCyc and MetNetDB virtually use the same number of EV codes, the mapping is framed considering four major EV codes. KEGG uses only one EV for pathways namely *'manually entered from published materials'*. The EV code for KEGG to *Direct* is mapped using the rules like;

If Di = BioCyc/AraCyc/MetaCyc, and EV = EV-Exp, then Change EV = Direct

Unification of the EV codes for the databases is based on the expert knowledge. EV code mapping is done with respect to a reference data source and unified according to the set of rules above.

Step 3. Confidence Weight (*CWi*) Assignment

Researchers typically have databases that they treat as favored sources for different types of information. Since there is no precise rule for deciding which database is more correct and up to date, a user defined score, a *confidence weight (CW)* is applied. The EV mapping process is interactive and provides flexibility in choice for databases. Confidence is defined as,

CWi = {Very Strong, Strong, Moderate, Poor, Very Poor}
For example: *CW **KEGG**: Strong, CW**BioCyc**: Normal*

Step 4. EVint (Rule Set-II)

Using heuristic rules, integrated EVcode is calculated.

Step 5. Decode *EVint* value

The EV value from Step 4 is decoded using:
$EVint = \Sigma \ (CWi^* \ EV)/ \mid i \mid = x$

Step 6. Rank Index

- Rank the journals in their order of importance.
- Make an ordered list of journals assigning *Rank*.
- Rank the conferences in order of their importance.
- Make an ordered list of conferences.
- *Assign:*
     If the publication in not in the list, Then, *Rank = low*
     Else, *Rank = as defined by the list*

Step 7. Value Factor *(VF)*

The *VF* measures support for the entity using the publication evidence. This is a quantitative index with a temporal function.

For t = current year, compute $VF\ (t) = |P\ (t\text{-}2)|\ /\ |P|$ where,

> $|P\ (t\text{-}2)|$ = Number of publications in the last $(t\text{-}2)$ years for Di, and
> $|P|$ = Total number of publications listed in Di.

Step 8. Reference Index *(RIint)*

> - Compute *RI* for *{D1,…Dn}* given by,
> $$RIi\{t\} = f\,\{Rank,\ VF\}$$
> - Compute *RIint for a* pathway as;
> $$RIint = max\ \{RIi\}$$

3.1. Integration models

Data integration aims to work with repositories of data from a variety of sources. As such, two databases may not provide identical information, and integrating these two databases may yield a richer resource for analysis. The conditions under which data is collected and the supporting references play a crucial role in making the analysis more meaningful. So far, the integration approaches have focused on different types of pathways. The same pathway can have different representations in different databases.

For example, a known pathway like Glycolysis is represented in different ways in KEGG and BioCyc as shown in Figure 3. A universal tool to integrate all types of pathways may not be a focus. Additionally, different databases employ various data representations that may not provide easy user access or user friendly. Figure 3(a) and 3(b) illustrate representational difference between two data sources for the same pathway. Various data integration models are defined below.

- *Syntactic Networks*: Syntactic networks adhere to the syntax of a set of words as given by the representation of the data and do not interpret the meaning associated. Syntactic heterogeneity is a result of differences in representation format of data.
- *Semantic Networks* (SN): Semantic heterogeneity is a result of differences in interpretation of the 'meaning' of data. Semantic models aim to achieve semantic interoperability, a dynamic computational capability to integrate and communicate both the explicit and implicit meanings of digital content without human intervention.
- Several features of SN make it particularly useful for integrating biological data include, ability to easily define an inheritance hierarchy between concepts in a network format, allow economic information storage and deductive reasoning, represent assertions and cause effect through abstract relationships, cluster related information for fast retrieval, and adapt to new information by dynamic modification of network structures [44]. An important feature of SN is the ease and speed to retrieve information concerning a particular concept. The use of semantic relationships ensures clustering together related concepts in a network. For example, protein synonyms, functional descriptions, coding sequences, interactions, experimental data or even relevant research articles can all be represented by semantic agents, each of which is directly linked to the corresponding protein agent.

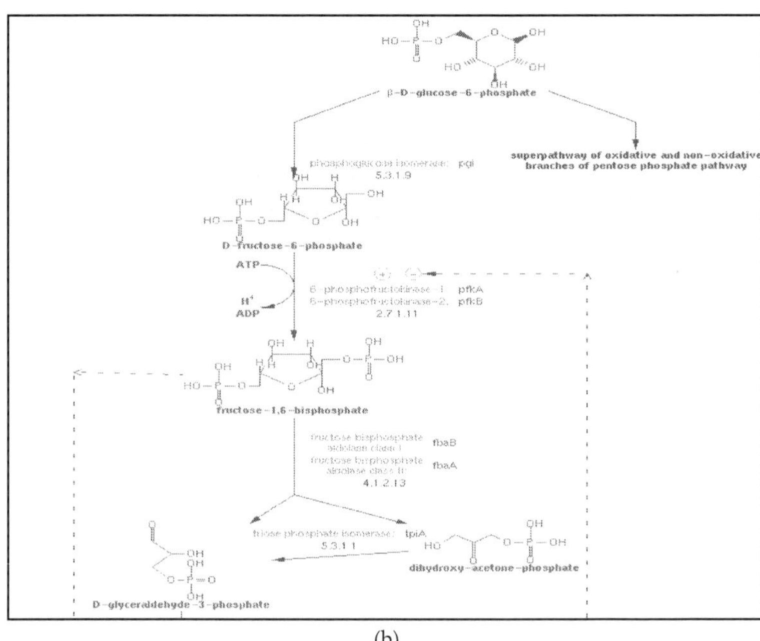

(a)

(b)

Figure 3. (a) Pathway from KEGG- Glycolysis (b) BioCyc- Glycolysis

Biological information can be retrieved effectively through simple relationship traversal starting from a query agent in the semantic network. Two approaches primarily in practice for SNs are;

1. memory-mapped data structure and
2. indexing flat files.

In the memory-mapped data structure approach, subsets of data from various sources are collected, normalized, and integrated in memory for quick access. While this approach performs actual data integration and addresses the problem of poor performance in the federated approach, it requires additional calls to traditional relational databases to integrate descriptive data. While data cleaning is being performed on some of the data sources, it is not being done across all sources or in the same place. This makes it difficult to quickly add new data sources. In the indexing flat files approach, flat text files are indexed and linked thus supporting fast query performance.

- *Causal Models*: A causal model is an abstract model that uses cause and effect logic to describe the behaviour of a system. Ex: Expression Quantitative Trait Loci: (eQTLs) eQTL analysis is to study the relationship between genome and transcriptome. Gene expression QTLs that contain the gene encoding the mRNA are distinguished from other transacting eQTLs. eQTL mapping tries to find genomic variation to explain expression traits. One difference between eQTL mapping and traditional QTL mapping is that, traditional mapping study focuses on one or a few traits, while in most of eQTL studies, thousands of expression traits get analyzed and thousands of QTLs are declared.
- *Context likelihood of relatedness* (CLR): It uses transcriptional profiles of an organism across a diverse set of conditions to systematically determine transcriptional regulatory interactions. *CLR* is an extension of the relevance network approach. (http://gardnerlab.bu.edu/software&tools.html). [34] Presented architecture for context-based information integration to solve semantic difference problem, defined some novel modeling primitives of translation ontology and propose an algorithm for translation.
- *Bayes Networks (BN)*: Probabilistic graphical models that represent a set of variables and their probabilistic independencies. For example, a BN could represent the probabilistic relationships between diseases and symptoms. Given symptoms, the network can be used to compute the probabilities of the presence of various diseases. Bayes networks focus on score-based structure inference. Available heuristic search strategies include simulated annealing and greedy hill-climbing, paired with evaluation of a single random local move or all local moves at each step. [45] Bases his approach on the well-studied statistical tool of Bayesian networks [46]. These networks represent the dependence structure between multiple interacting quantities (e.g., expression levels of different genes). His approach, probabilistic in nature, is capable of handling noise and estimating the confidence in the different features of the network.
- *Hidden Markov Models (HMM)*: HMM is a statistical model that assumes the system being modeled to be a Markov process with unknown parameters, and determines the

hidden parameters from the observable parameters. The extracted model parameters can then be used to perform further analysis, for example for pattern recognition applications. An HMM can be considered as the simplest dynamic Bayesian network. HMMs are being applied to the analysis of biological sequences, in particular DNA since 1998 [47].

3.2. Need to use open grid service architecture ogsa-dai for data access and integration

Apart from the ubiquitous call for more functionality, bioinformatics projects with commercial users/partners are very anxious about the security of their data. The issue is further complicated by the lack of coherent security models with the evolving WS-RF and WS-I specifications which OGSA-DAI now supports. This issue needs to be resolved if bioinformatics projects with commercial users/partners are not to be deterred from adopting the product despite its utility. In contrast to the diversity of its data resources, a limited range of operations on these resources is typically required. For instance, one operation is to create a study data set by aggregating data from iterative searches of remote data collections using the same taxonomy object (representing a species or other group) as the search parameter [48].

3.3. Handling the heterogeneity in data representation among databases

For biological plant pathways, various databases incorporate information about an entity/reaction/pathway to a level of detail and define their own data format. This includes information like number of fields, column label/tag, pathway name(s), etc. At the outset, common information across the tables may look limited and hard to extract mainly because of the tag or synonyms (other names) of pathway. Before proceeding for integration of a pathway across data sources following steps need to be carried out. For biological pathway integration, following needs to be considered.

- What is the aim of integration?
  To query autonomous and heterogeneous data sources through a common, uniform schema (TARGET SCHEMA).
- How will the integrated data be used?
  - Resolving various conflicts between source and target schema.
  - Offering a common interface to access integrated information.
  - Preserving the autonomy of participating systems.
  - Easily integrating data sources without major modification.
- Is it within a single data source or across sources?
- Does it support web based integration?
- Does it encompass the dynamic nature of the data?
- What are the data, source, user models, and assumptions underlying the design of integration system?

Specific data integration problems in the biological field include:

- Some biological data sources do not provide an expressive language
- Derived wrapper (operate in two modes)
  - Traditional wrapper
  - Virtual source that buffers the execution result of a local application
- Data Model Inconsistencies requires complex data transformation coding
- Data Schema Inconsistencies
- Schema matching: error-prone task
- Mapping info: systematically managed
- Domain Expert participation
- Along with the data schema consistencies there may be data level inconsistencies such as:
- Data conflict as each object has its own data type, and may be represented in different formats
- Different Query Capabilities affect the query optimization of data integration system
- Miscellaneous: Network environment, Security

File formats: For biological pathways, various data sources incorporate information about an entity/reaction/pathway to a level of detail and define their own data format. This includes information like number of fields, column label/tag, pathway name(s), etc. At the outset, common information across the tables may look limited and hard to extract mainly because of the tag or synonyms (other names) of pathway. One of the other important differences in the way these data sources are developed lies in the synonym representations. Some of the data sources limit the synonyms to 10 others may not result into may be over 40 synonyms. While we look at the data integration mechanism, if the names of the compounds do not match, then the search should be carried forward with the list of synonyms. In integrating different data bases this will take different search time. Also, since the field names (compound names) did not match, the search must unify the field names and generate a new list of synonyms.

Granularity of information: Different pathway databases may model pathway data with different levels of details. This primarily depends on the process definition. For example, one database might treat processes together as a single process, while another database might treat these as separate processes. Also, one database might include specific steps to be part of the process, while another database might not consider these steps. Additionally, the levels of details associated with a certain data base necessitate pathway data modeling with different levels of granularity. Different pathway data formats (e.g., SBML and BIND XML) have been used to represent data with different levels of details. A semantic net based approach to data integration is proposed in [49].

Heterogeneous formats: As the eXtensible Markup Language (XML) has become the lingua franca for representing different types of biological data, there has been a proliferation of semantically-overlapping XML formats that are used to represent diverse types of pathway data. Examples include the XML-derivatives KGML, SBML, CellML, PSI MI, BIND XML, and Genome Object Net XML. Efforts have been underway to translate between these

- ***SynList** {entity name} = **SynList** {v^{1j}_{kn}}*
- *EV^{ij}_k = {EV^{ij}_{k1}, …. EV^{ij}_{kh}} set of 'h' EV Codes for {s^i, p^{ij} , d^{ij}_k }, for example;*
  - *EV^{1j}_1= {Set of EVcodes given by Biocyc for E.coli for TCA cycle}*
  - *EV^{1j}_2= {Set of EV codes given by KEGG for E.coli for TCA cycle}*
  - *EV^{2j}_3= {Set of EV codes given by AraCyc for Arabidopsis for TCA cycle}*
  - *EV^{2j}_2= {Set of EV codes given by KEGG for Arabidopsis for TCA cycle}*
- *RI^{ij}_k: Reference index for a database d^{ij}_k*
- *RI^{ij}_{int}: Reference index for the integrated pathway*
- *CW^{ij}_k: Confidence weight for a database d^{ij}_k*
- *CW^{ij}_{int}: Confidence weight of the integrated pathway p^{ij} within a species*
- *V^{ij}_{int}: Integrated node table for a species S^i, for a pathway p^{ij}*
- *E^{ij}_{int}: Integrated edge table for a species S^i, for a pathway p^{ij}*
- *(v^{1j}_{kn}, e^{ij}_{km}) = (node 'n', edge 'm') in d^{1j}_k of s^1 for p^{1j};*
- *ATT {(v^{1j}_{kn} ,(A)}= {v^{1j}_{kn}, (A_1, A_2, A_3, A_4, …A_s)} = set of attributes of the node v^{1j}_{kn}*
- *ATT {(e^{ij}_{km}, (B)} ={(e^{ij}_{km}, (B_1, B_2, B_3, … B_t)} = set of attributes of edge e^{ij}_{km}*
- *$DATT$ {v^{1j}_{kn} ,(δA)} = set of derived attributes of the node v^{1j}_{kn} (EV_i, CW_i, RI_i)*
- *$DATT$ {e^{1j}_{kn} ,(δB)}= set of derived attributes of the edge e^{1j}_{kn} (EV_i, CW_i, RI_i)*
- *δV^{ij}_k = Set of derived node attributes for Integrated pathway {EV_{int}, CW_{int}, RI_{int}}*
- *δE^{ij}_k= Set of derived edge attributes for Integrated pathway {EV_{int}, CW_{int}, RI_{int}}*
- *V^{ij}_{int} = {Σ V^{ij}_k } $_{for\ k=1\ to\ n}$*
- *E^{ij}_{int}= {Σ E^{ij}_k} $_{for\ k=1\ to\ n}$*
- *P^{ij}_{int} = Integrated pathway from multiple DSs = {Σ P^{ij}_k } $_{for\ k=1\ to\ n}$*

4.2. Biological Pathway Data Integration Algorithm

Following selections and inputs are defined by the user.

- User selected inputs: Species, Pathway, Data sources/database
- User inputs: Confidence assigned to each database
- User defined filters (**UDF**) for entities like substrate nodes, H_2O, CO_2 etc. for integrated pathway [P^{ij}_{int} = G (V_{int}, E_{int})],

Step 1.

   For each user selected pathway ***P^{ij}** for a species s^i*

List D^{ij} *(d^{1j}_1,… d^{nj}_k),*   ***(KEGG, BioCyc, MetNetDB etc)****

Step 2._**Define** rules to classify the interactions, for example;

 - If the pathway is *signal transduction,* then use the *classifier (Table 1)*for interactions
 - If the pathway is *metabolic,* then *reaction* is a general representation of the interaction

  Sort *(d^{1j}_1,… d^{nj}_k)* according to species *(s^i,d^{ij}_1), (s^j,d^{ji}_1)* etc.

  Generate a set of (nodes, edges) from all the input data sources *{(V^{ij}, E^{ij})} = {(V^{ij}_1, E^{ij}_1), (V^{ij}_2, E^{ij}_2)….. (V^{ij}_s, E^{ij}_s)}*

where, $V^{ij}{}_k = \{v^{ij}{}_{k1}, v^{ij}{}_{k2}, ...v^{ij}{}_{kt}\}$ and $E^{ij}{}_1 = \{e^{ij}{}_{k1}, e^{ij}{}_{k2},....e^{ij}{}_{ku}\}$

Step 3.

For $k = 1, ..., q\ (d^{1j}{}_1,... d^{1j}{}_k)$,

   For $s = 1,.., n$, and $q = 1, ...m$,

        List *ATT $\{(v^{1j}{}_{ks}, (A)\}$*

        List *ATT $\{(e^{ij}{}_{kq}, (B)\}$*

        Select *$v^{ij}{}_{k1} \in V^{ij}{}_k\ C\ d^{1j}{}_k$*

        For all *$p = 1\ to\ n$*

        Check *for $v^{ij}{}_{k,1} \in V^{ij}{}_p$* (node name match across data sources)

        If *YES*, then **Apply** *EV* integration algorithm

            Generate **DATT** *$\{v^{1j}{}_{kn}, (\delta A)\}$,* **DATT** *$\{e^{1j}{}_{kn}, (\delta B)\}$,*

     Else, *For $p = 1\ to\ n$,*

            For $t = 1, z$

            Check if *$v^{ij}{}_{k,1} \in$* **SynList** *$\{v^{ij}{}_{p,t}\}$* (node name(A) with Synlist(B))

            If YES, then Apply EV integration algorithm,

            Generate **DATT** *$\{v^{1j}{}_{kn}, (\delta A)\}$,* **DATT** *$\{e^{1j}{}_{kn}, (\delta B)\}$,*

    Else,

            Check if **SynList** *$\{v^{ij}{}_{k,1}\}$* has a match with *$v^{ij}{}_{p,t}$*

            If YES, then Apply EV integration Algorithm

    Else,

            Check if **SynList** *$\{v^{ij}{}_{k,1}\}$* has a match with **SynList** *$\{v^{ij}{}_{p,t}\}$*

            If $v^{ij}{}_{k,1} = v^{ij}{}_{p,l}$ is TRUE,

            Then,

            Include *$v^{ij}{}_{k,1}$* with the matched node name *$v^{ij}{}_{k-1,p} \in V^{ij}{}_{k-1}$*

            Compute *$(\delta V^{ij}{}_k, \delta E^{ij}{}_k)$*

This is the node name for the integrated database for the species. **Level 1**

        Generate SynListInt = $\{$**SynList** $(v^{ij}{}_{k,1})$ U *Synlist* $(v^{ij}{}_{p,l})$U...$\}$ without duplication

        Associate DOI (date of integration)

        Generate *$P^{ij}{}_{int}$*

$P^{ij}{}_{int} = \{\Sigma\ P^{ij}{}_k\}$ *for k=1 to n* $= [\{\ V^{ij}{}_{int}, E^{ij}{}_{int}\} + \{\ \Sigma\ \delta V^{ij}{}_{kt}, \Sigma\ \delta E^{ij}{}_k\}$ *for k=1 to n* $]$ *at t= t1*

    $= \Sigma\ \{ATT\ [(v^{1j}{}_{kn}, (A)]\}, ATT\ [(e^{ij}{}_{km}, (B)]\} + \Sigma\ \{DATT\ \{v^{1j}{}_{kn}, (\delta A), DATT\ \{e^{1j}{}_{kn}, (\delta B)\}$ *for*

all n, m $\{\delta V^{ij}{}_k\ \delta E^{ij}{}_k\}$

Step 4.

    Repeat *Step 2-3 for $e^{ij}{}_k \in E^{ij}{}_k$ in $(d^{ij}{}_1,... d^{ij}{}_k)$, for p^{ij}*

    Include information associated with the edge, as given by 'edges' such as reaction, enzyme, by products and substrates along with attributes like evidence, reference publications, context etc.

** Outputs $E^{ij}{}_{int}$ table for s^i using $(d^{1j}{}_1,... d^{1j}{}_k)$, with $EV^{ij}{}_{int}$, $CW^{ij}{}_{int}$ and $RI^{ij}{}_{int}$. **Level 1.** **

Step 5.

    Generate integrated pathway by consolidating outputs $G(V^{ij}_{int}, E^{ij}_{int})$ for s^i

Step 6.

For i =1,...n
    Repeat steps 2- 4 to integrate P^{ij} for all species s^i
** **This generates Table** $(V^j_{int}, E^j_{int}) = \{(V^{ij}_{int}, E^{ij}_{int})$ U $(V^{jj}_{int}, E^{jj}_{int})U...\}$ for S^i, **for all** $i =1, ..n)$, for a p^{ij}. **Level 2***

Step 7.

For (j = 1,....p)
    Integrate for all P^{ij}
This **generates output table $(V_{int}, E_{int}) = (V^j_{int}, E^j_{int})$ U (V^k_{int}, E^k_{int}) U....for all $(j = 1,....p)$. **Level 3***

Step 8.

    Apply UDF (User defined filter)

5. Querying Integrated Pathway

Once the data integration is accomplished, extracting information from the integrated data will be of interest to the biologist. There are various mechanisms to extract information from the integrated database generated. Some of these are described below.

Granular computing with semantic network structure captures the abstraction and incompleteness associated with biological plant pathway data. It is inspired by the ways in which humans granulate information and reason with coarse grained information. The three basic concepts underlying the human cognition are granulation, organization, and causation. Granulation involves decomposition of whole into parts, organization involves integration of parts into whole, and causation involves associations of cause and effects. The fundamental issues with granular computing are granulation of the universe, description of granules, and relationships between granules. The basic ideas of crisp information granulation have appeared in related fields, such as interval analysis, quantization, rough set theory, Demster Shafer theory of belief functions, divide and conquer, cluster analysis, machine learning, data bases and many others. Granules may be induced as a result of 1) equivalence of attribute values, 2) similarity of attribute values, and define the granules 3) equality of attribute value. We use granules for defining the user queries associated with the integrated pathway. Based on user (biologist) choice, granules can be defined to view the integrated pathway. This provides flexibility to the biologist for using the information.

Previous approaches towards metabolic network reconstruction have used various algorithmic methods such as name-matching in IdentiCS [52] and using EC-codes in metaSHARK [53] to link metabolic information to genes. The AUtomatic Transfer by Orthology of Gene Reaction Associations for Pathway Heuristics (AUTOGRAPH) method

[54] uses manually curated metabolic networks, orthologue and their related reactions to compare predicted gene-reaction associations.

Arrendondo [55] Proposes to develop a process for the continuous improvement of the inference system used, which is applicable to any such data mining application. It involves the comparison of several classifiers like Support Vector Machines (SVMs), Human Expert generated Fuzzy, and Genetic Algorithm (GA) generated Fuzzy and Neural Networks using various different training data models. In his approach, all classifiers were trained and tested with four different data sets: three biological and a synthetically generated mixture data set. The obtained results showed a highly accurate prediction capability with the mixture data set providing some of the best and most reliable results.

6. Conclusion

Biological database integration is a challenging task as the databases are created all over the world and updated frequently. For biological data sources that may be derived from an earlier existing data source, it is also important to identify the evidence of the data source represented by the evidence code, to be included as a candidate for integration. In most data integration algorithms the user does not participate thus leading to an integrated data source with any effective utility towards analysis.

Large scale integration of pathway databases promises to help biologists gain insight into the deep biological context of a pathway. In this chapter, we presented algorithms that help user to select their choice of data sources and apply Evidence code algorithm to compute an integrated EV code and RI for the pathway data of interest. The ultimate goal is to generate a large-scale composite database containing the entire metabolic network for an organism. This qualitative approach includes aspects like user confidence scores for databases for mapping EV and generating RI for a given pathway. For the TCA pathway results show that generating such a mapping is helpful in visualizing the integrated database that highlights the common entities as well as the specifics of each database. As the database confidence weight selection is user specific, the integration yields different results for different users for the same database which will allow users to explore the effects of different hypotheses on the overall network. Once the integrated evidence code is generated, then data integration algorithm is applied to get the integrated pathway data. To best attempt integration of such data it is imperative to include user participation as user mostly identifies the associations and behavior of various compounds, reactions, genes in a given biological pathway leading to significant diagnosis.

Author details

Shubhalaxmi Kher
Electrical Engineering, Arkansas State University, USA

Jianling Peng
Samuel Roberts Noble Foundation, USA

Eve Syrkin Wurtele
Department of Genetics, Development and Cell Biology, Iowa State University, USA

Julie Dickerson
Electrical and Computer Engineering, Iowa State University, USA

7. References

[1] Akula, S.; Miriyala, R.; Thota, H.; Rao, A.; Gedela, S. Techniques for Integrating –omics Data, Bioinformation, Views and Challenges, 2009.

[2] Saccharomyces genome database. http://www.yeastgenome.org/

[3] KEGG: Kyoto Encyclopedia of Genes and Genomes. http://www.genome.jp/kegg/

[4] TAIR- AraCyc: http://www.arabidopsis.org/biocyc/

[5] Thimm, O; Blasing, O; Gibon, Y; Nagel, A; Meyer, S; Kruger, P; Selbig, J; Muller, L; Rhee, S; and Stitt, M. MAPMAN: a user driven tool to display genomics data sets onto diagrams of metabolic pathways and other biological processes, The Plant journal (2004) 37, pp 914-939. http://www.uky.edu/~aghunt00/PLS620/papers.htm/ Systems%20approaches%20copy/MAPMAN.pdf

[6] BIND: Biomolecular Interaction Network Database http://metadatabase.org/wiki/BIND_-_Biomolecular_Interaction_Network_Database

[7] Bajic VB, Veronika M, Veladandi PS, Meka A, Heng MW, Rajaraman K, Pan H, Swarup S. Dragon Plant Biology Explorer. A text-mining tool for integrating associations between genetic and biochemical entities with genome annotation and biochemical terms list, Plant Physiol. 2005 Aug; 138(4):1914-25.

[8] Pandey R, Guru R K, Mount D W. Pathway Miner: extracting gene association networks from molecular pathways for predicting the biological significance of gene expression microarray data, Bioinformatics. 2004 Sep 1;20(13):2156-8. Epub 2004 May 14.

[9] RegulonDB database: Escheichia Coli k-12 transcriptional network. http://regulondb.ccg.unam.mx/

[10] PlantCare a database. http://bioinformatics.psb.ugent.be/webtools/plantcare/html/

[11] PLACE: a database of Plant Cis-acting regulatory netowrk. http://www.dna.affrc.go.jp/PLACE/

[12] EPD: Eukaryotic promoter database. http://epd.vital-it.ch/

[13] TRRD: transcription regulatory regions database http://wwwmgs.bionet.nsc.ru/mgs/gnw/trrd/

[14] Athamap: http://www.athamap.de/

[15] TRANSFAC: http://www.gene-regulation.com/pub/databases.html/

[16] Friedman N, Linial, M; Nachman,I; and Pe'er, D. Using Bayesian Networks to Analyze Expression Data, Journal of computational biology, Volume 7, Numbers 3/4, 2000, pp. 601-620.

[17] Schadt, et.al, *An Integrative Genomics Approach to Infer Causal Associations Between Gene Expression and Disease, Nature Genetics, vol.37, number 7, July 2005, pp, 710-717.*

[18] The EMBL Nucleotide Sequence Database (http://www.ebi.ac.uk/embl/).

[19] Liu, Y.; Wang, Y.; Liu, Y.;, Tan, Z. Data Integration of Bioinformatics Database Based on Web Services, International Journal of Web Applications, Volume 1, Number 3, 2009.

[20] UCLA-DOE Institute for Genomics and Proteomics. http://dip.doe-mbi.ucla.edu/dip/Main.cgi

[21] IntAct: open source database system and analysis tools for molecular interaction data.http://www.ebi.ac.uk/intact/

[22] GRID: http://www.moldiscovery.com/soft_grid.php/

[23] Zanzoni, A. Montecchi-Palazzi, L. Quondam, M. Ausiello, G. Helmer-Citterich, M. Cesareni, G. *MINT: A Molecular INTeraction database. Elsevier FEBS Letters, 2002, Volume 513, Issue 1, Pages 135-140.*

[24] Coessens B. et.al, *INCLUSive: A Web Portal and Service Registry for Microarray and Regulatory Sequence Analysis, Nucleic AciDS research, 2003, vol. 31, No.13. pp. 3468-3470.* http://tomcatbackup.esat.kuleuven.be/inclusive/

[25] Achard, F.; Vaysseix, G.; Barillot, E. XML, Bioinformatics, and Data Integration, Bioinformatics Review, Evry, France, 2001, pp. 115-125.

[26] Pathway Data List. http://cbio.mskcc.org/prl

[27] Hsing, M., Cherkasov, A. Integration of Biological Data with Semantic Networks, Current Bioinformatics, 2006, 1 000-000.

[28] Chung, M., Lim, M., Bae, M., Park, S. Customized Biological Database Integration for cDNA Microarray, RECOMB 2005, Research in Computational and Molecular Biology, Cambridge, 2005.

[29] Gopalcharyulu, P. Lindfors, E. et.al. Data integration and visualization system for enabling conceptual biology, BioInformatics, Vol.21, Suppl 1 2005, pp. i177-i185.

[30] Rzhetsky, A et.al, GeneWays: A System for Extracting, Analyzing, Visualizing and Integrating Molecular Pathway Data, Journal of Bioinformatics, 2004, 43-53.

[31] Zucker, J.,Luciano, J., Brandes, A. Lin, X. Semantic Aggregation Integration and Inference: Three case studies, ISMB 2005.

[32] Hu, Z., Mellor, J., Wu, J., Yamada, T., Holloway, D., DeLisi, C. VisANT: Data integrating visual framework for biological networks and modules, Nucleic AciDS research, 2005 vol. 33.

[33] Zhang Z.; Bajic, V.; Yu, J.; Cheung, K.; Townsend, J. Data Integration in Bioinformatics: Current Efforts and Challenges. Bioinformatics: Trends and Methodologies, Intech, 2011.

[34] Zhang, D. and Jing, L., *Context based Numerical information, IEEE conference on E-commerce Technology 2005*Arredondo, T., Seeger, M., Dombrovskaia, L., Avarias, J., Calderón, F., Candel, D., Muñoz, F., Latorre, V., Agulló, L., Cordova, M., and Gómez, L.: "Bioinformatics Integration Framework for Metabolic Pathway Data-Mining". In: Ali, M., Dapoigny, R. (eds): Innovations in Applied Artificial Intelligence. Lecture Notes in Artificial Intelligence, Vol. 4031. Springer-Verlag, Berlin (2006) pp. 917-926.

[35] PATIKA: http://www.iam.metu.edu.tr/research/groups/compbio/PATIKA_METU04.pdf

[36] INHO: http://www.inoh.org/

[37] TRANSPATH. http://www.ncbi.nlm.nih.gov/pubmed/12519957

[38] ReactomeSTKE. http://stke.sciencemag.org/

[39] MetaCyc. http://metacyc.org/

[40] Kher, S; Jianling Peng; SyrkinWurtele, E.; Dickerson, J. A Symbolic computing approach to evidence code mapping for biological data integration and subjective analysis for reference associations for metabolic pathways, Annual Meeting of the North American Fuzzy Information Processing Society, 2008, NAFIPS 2008. NY 2008. pp. 1-6.

[41] Kher, S; Dickerson, J; Rawat N. Biological pathway data integration trends, techniques, issues and challenges: A survey, Nature and biologically inspired computing, NaBIC 2010, Second World Congress, Fukuoka, Japan, 2010, pp.177 – 182.

[42] MetNetDB. http://www.metnetdb.org/MetNet_db.htm

[43] Karp, P. D., Paley, S., Krieger, C. J. An Evidence Ontology for Use in Pathway/Genome DS, Pacific Symposium on Biocomputing 2004, pp. 190-201, Singapore Bounsaythip, C., Lindfors, E., Gopalacharyulu, P., Hollmen, J., and Oresic, M. *Network Based Representation of Biological Data for Enabling Context Based Mining, Bioinformatics, vol.21, suppl 1. 2005, pp. 177-185.*

[44] Newman, M. E. J and Leicht, E. A *Mixture Models and Exploratory Analysis in Networks, Physics, May 2007.*

[45] Pearl. J. (2000) *Causality: Models, Reasoning, and Inference.*Cambridge University Press, 2000.

[46] Pearl, J. (1988). *Probabilistic Reasoning in Intelligent Systems: Networks of Plausible Inference.* San Mateo, CA, USA: Morgan Kaufmann Publishers.

[47] Christopher Nemeth, STOR-I, Hidden Markov Models with Applications to DNA Sequence Analysis.

[48] Crompton, S.; Matthews, B.; Gray, A.; Jones, A.; White, R. Data Integration in Bioinformatics Using OGSA-DAI, In Proceedings of Fourth All Hands Meeting, 2005.

[49] Cheung Kei-hoi; Qi, P; Tuck,D; Krauthammer,M. A Semantic Web Approach to Biological Pathway Data Reasoning and Integration, Elsevier Vol. 4, issue 3, Sep 2006, pp. 207-215.

[50] RDF-OWL. http://www.w3.org/RDF/

[51] BioPAX: http://www.biopax.org/

[52] Sun, J. and Zeng, A. , IdentiCS – Identification of coding sequence and *in silico* reconstruction of the metabolic network directly from unannotated low-coverage bacterial genome sequence, *BMC Bioinformatics* 2004, 5:112 doi:10.1186/1471-2105-5-112

[53] Pinney, J.W., Shirley, M.W., McConkey, G.A., Westhead, D.R. (2005) MetaSHARK: software for automated metabolic network prediction from DNA sequence and is application to the genomes of *Plasmodium falciparum* and *Eimeria tenella, Nucleic Acids Research*, 33, 1399-1409.

[54] Notebaart, R. A., F. H. van Enckevort, C. Francke, R. J. Siezen, and B. Teusink. 2006. Accelerating the reconstruction of genome-scale metabolic networks. BMC Bioinformatics 7:296

[55] Arredondo, T., Seeger, M., Dombrovskaia, L., Avarias, J., Calderón, F., Candel, D., Muñoz, F., Latorre, V., Agulló, L., Cordova, M., and Gómez, L.: "Bioinformatics Integration Framework for Metabolic Pathway Data-Mining". In: Ali, M., Dapoigny, R.(eds): Innovations in Applied Artificial Intelligence. Lecture Notes in Artificial Intelligence, Vol. 4031. Springer-Verlag, Berlin (2006) p. 917-926

Investigation on Nuclear Transport of *Trypanosoma brucei*: An *in silico* Approach

Mohd Fakharul Zaman Raja Yahya,
Umi Marshida Abdul Hamid and Farida Zuraina Mohd Yusof

Additional information is available at the end of the chapter

http://dx.doi.org/10.5772/47448

1. Introduction

1.1. Trypanosomiasis

A group of animal and human diseases caused by parasitic protozoan trypanosomes is called trypanosomiases. The final decade of the 20th century witnessed a frightening revival in sleeping sickness (human African trypanosomiasis) in sub-Saharan Africa. Meanwhile, Chagas' disease (American trypanosomiasis) remains one of the most widespread infectious diseases in South and Central America. Arthropod vectors are responsible for the spread of African and American trypanosomiases, and disease restraint through insect control programs is an attainable target. However, the existing drugs for both illnesses are far from ideal. The trypanosomes are some of the earliest diverging members of the Eukaryotae and share several biochemical oddities that have inspired research into discovery of new drug targets. Nevertheless, discrepancies in mode of interactions between trypanosome species and their hosts have spoiled efforts to design drugs effective against both species. Heightened awareness of these neglected diseases might result in progress towards control through increased financial support for drug development and vector eradication [1].

Trypanosome is a group of unicellular parasitic flagellate protozoa which mostly infects the vertebrate genera. A number of trypanosome species cause important veterinary diseases, but only two cause significant human diseases. In sub-Saharan Africa, *Trypanosoma brucei* causes sleeping sickness or human African trypanosomiasis whilst in America, *Trypanosoma cruzi* causes Chagas' disease (Figure 1) [2]. Meanwhile, the life cycle of these parasitic protozoa engage insect vectors and mammalian hosts (Figure 2) [1]. All trypanosomes require more than one obligatory host to complete their life cycle and are transmitted via vectors. Most of the species are transmitted by blood-feeding invertebrates, however there

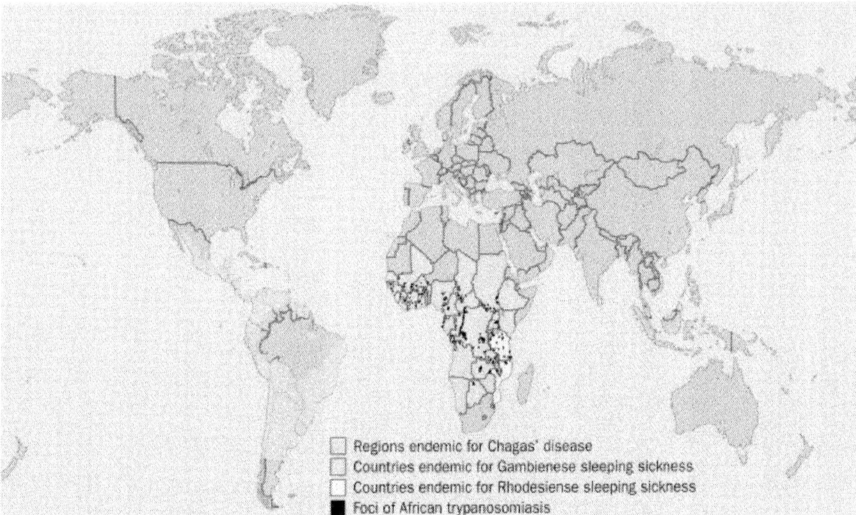

Figure 1. Geographic distribution of *Trypanosoma brucei* and *Trypanosoma cruzi*, showing endemic countries harboring these diseases [2].

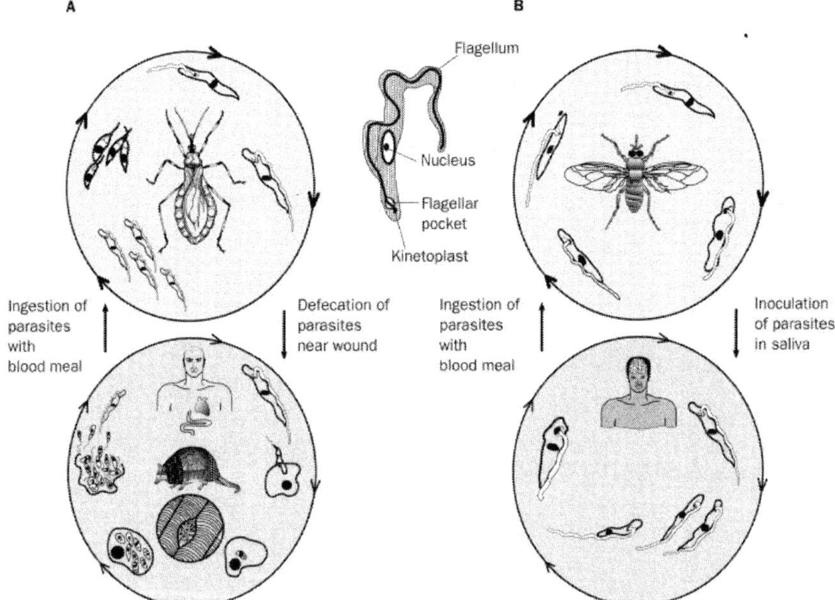

Figure 2. Life cycles of (A) *Trypanosoma cruzi* and (B) *Trypanosoma brucei*. Upper cycles represent different stages that take place in the insect vectors. Lower cycles represent different stages in man and other mammalian hosts [1].

are distinct mechanisms among the varying species. In the invertebrate hosts they are generally found in the intestines as opposed to the bloodstream or any other intracellular environment in the mammalian host. As trypanosomes develop through their life cycle, they undergo a series of morphological changes [3] as is typical of trypanosomatids.

The life cycle often consists of the trypomastigote form in the vertebrate host and the trypomastigote or promastigote form in the gut of the invertebrate host. Intracellular lifecycle stages are normally found in the amastigote form. The trypomastigote morphology is unique to species in the genus Trypanosoma.

The genome organization of *T. brucei* is splitted into nuclear and mitochondrial genomes. The nuclear genome of *T. brucei* is made up of three classes of chromosomes according to their size on pulsed-field gel electrophoresis, large chromosomes (1 to 6 megabase pairs), intermediate chromosomes (200 to 500 kilobase pairs) and mini chromosomes (50 to 100 kilobase pairs) [4]. The large chromosomes contain most genes, while the small chromosomes tend to carry genes involved in antigenic variation, including the variant surface glycoprotein (VSG) genes. Meanwhile, the mitochondrial genome of the Trypanosoma, as well as of other kinetoplastids, known as the kinetoplast, is characterized by a highly complex series of catenated circles and minicircles and requires a cohort of proteins for organisation during cell division. The genome of *T. brucei* has been completely sequenced and is now available online [5].

1.2. Nuclear transport

Nuclear transport of proteins and ribonucleic acids (RNAs) between the nucleus and cytoplasm is a key mechanism in eukaryotic cells [6]. The transport between the nucleus and cytoplasm involves primarily three classes of macromolecules: substrates, adaptors, and receptors. The transport complex is formed when the substrates bind to an import or an export receptor. Some transport substrates require one or more adaptors to mediate formation of a transport complex. Once assembled, these transport complexes are transferred in one direction across the nuclear envelope via aqueous channels that are part of the nuclear pore complexes (NPCs). Following dissociation of the transport complex, both adaptors and receptors are recycled through the NPC to allow another round of transport to occur. Directionality of either import or export therefore depends on the formation of receptor-substrate complex on one side of the nuclear envelope and the dissociation of the complex on the other. The Ran GTPase is vital in producing this asymmetry. Modulation of nuclear transport generally involves specific inhibition of the formation of a transport complex, however, more global forms of regulation also occur [7]. The general concept of import and export process is shown in Figure 3 [8].

1.3. *In silico* approach

In silico study is defined as an analysis which is performed using computer or via computer simulation. It involves the strategy of managing, mining, integrating, and interpreting

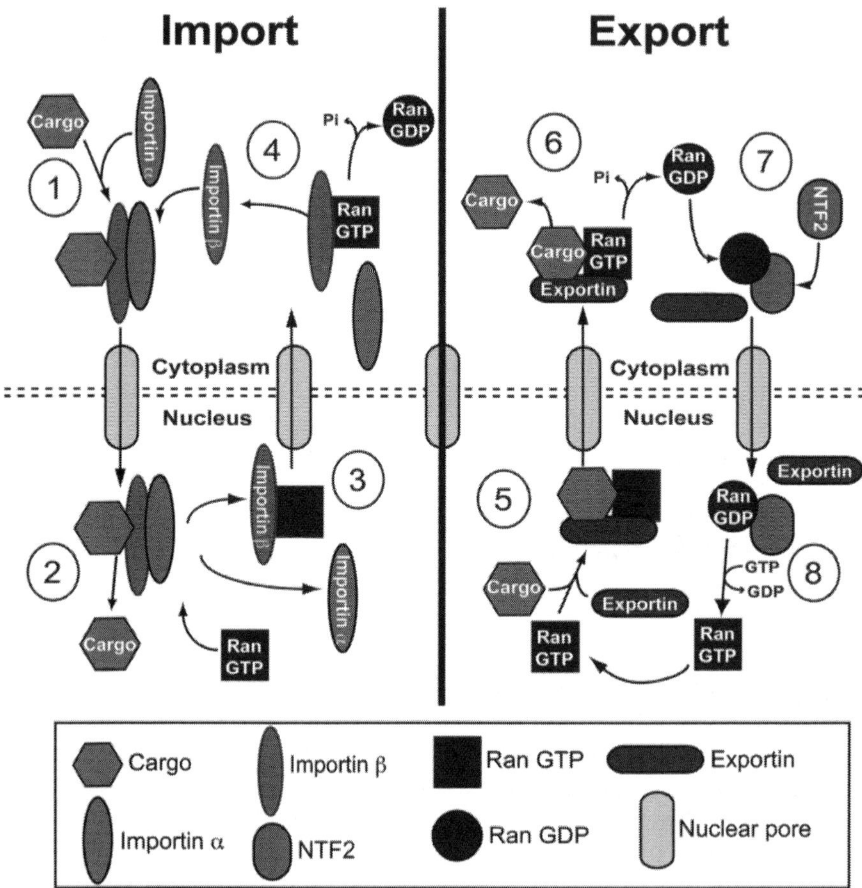

Key:

GTP   Guanine triphosphate
GDP   Guanine diphosphate
NTF2   Nuclear transport factor 2
RCC1   Regulator of chromosome condensation 1

Figure 3. For import of molecules, cytoplasmic cargo is identified by Importin *a*, which then binds to Importin b (1). This ternary complex translocates through the nuclear membrane and into the nucleus. Once there, RanGTP binds to Importin b and causes a dissociation of the complex, which releases cargo to the nucleus (2). Import receptors are then recycled back to the nucleus (3) through binding of RanGTP and export to the cytosol. RanGTP is then hydrolyzed to the GDP-bound state and causes the release of the import receptors (4) and the cycle starts over again. Export of cargo undergoes a similar mechanism. Exported molecules will bind to the export receptor with RanGTP and exit the nucleus (5). Next RanGTP is hydrolyzed to cause release of cargo into the cytoplasm (6). NTF2 specifically identifies RanGDP and returns it to the nucleus (7) for RCC1 to then exchange it to RanGTP (8) [8].

information from biological data at the genomic, metabalomic, proteomic, phylogenetic, cellular, or whole organism levels. The bioinformatics instruments and skills become crucial for *in silico* research as genome sequencing projects have resulted in an exponential growth in protein and nucleic acid sequence databases. Interaction among genes that gives rise to multiprotein functionality generates more data and complexity. *In silico* approach in medicine is not only reducing the need for expensive lab work and clinical trials but also is possible to speed the rate of drug discovery. In 2010, for example, researchers found potential inhibitors to an enzyme associated with cancer activity *in silico* using the protein docking algorithm EADock [9]. About 50 % of the molecules were later shown to be active inhibitors *in vitro* [9]. A unique advantage of the *in silico* approach is its worldwide accessibility. In some cases, having internet access or even just a computer is sufficient enough. Laboratory experiments either *in vivo* or *in vitro* both require more materials. In protein sequence analysis, *in silico* approach gives highly reproducible results in many cases or even exactly the same results because it only relies on comparison of the query sequence to a database of previously annotated sequences. However, in sophisticated analysis such as development of the 3-D structure of proteins from their primary sequences, discrepancies in results are to be expected due to the manual optimization which must consider several crucial steps such as template selection, target-template alignment, model construction and model evaluation.

1.4. Problem statements

Considering the importance of nuclear shuttling in many cellular processes, proteins responsible for the nuclear transport are vital for parasite survival. The presence of nuclear transport machinery was highlighted in the eukaryotic parasites such as *Plasmodium falciparum*, *Toxoplasma gondii* and *Cryptosporidium parvum*. However, the nuclear transport in *T. brucei* has not been established. Nuclear shuttling is one of the overlooked aspects of drug design and delivery. Exploitation of macromolecules movement across the nuclear envelope promises to be an exciting area of drug development. Furthermore, the divergence between host and parasite systems is always exploited as a strategy in drug development. Therefore, the exploitation of peculiarities of *T. brucei* nuclear transport machinery as compared to its host might be a promising strategy for the control of trypanosomiasis, which remains to be further investigated.

1.5. Objectives

This study is carried out to investigate the nuclear transport constituents of *T. brucei* by determining the functional characteristics of the parasite proteins. This includes functional protein domain, post translational modification sites and protein-protein interaction. The parasite proteins identified to exhibit the relevant functional protein domains, post translational modification sites and protein-protein interaction, are predicted as the true components for nuclear transport mechanism. This study also aims to evaluate the unique characteristics of proteins responsible for nuclear transport machinery between the parasites and human by determining the degree of protein sequence similarity. The information on the sequence level divergence between *T. brucei* proteins and their human counterparts may provide an insight into drug target discovery.

2. Materials and methods

Our *in silico* analyses were carried out using the public databases and web based programs (Table 1). The programs were employed to identify and annotate the parasite proteins involved in the nuclear transport mechanism. The identified parasite proteins were then compared with the human counterparts.

Analysis	Programme name	URL and Reference where available
Protein sequence retrieval	National Centre for Biotechnology Information (NCBI)	www.ncbi.nlm.nih.gov/
	Universal Protein Knowledgebase/SwissProt (UniProtKB/ SwissProt)	http://www.uniprot.org/
	TriTrypDB	http:// tritrypdb.org/ tritrypdb/
Clustering of protein sequences	BLASTClust	www.vardb.org/vardb/analysis/blastclust.html
Identification of protein domains	Conserved Domain Database (CDD)	http://www.ncbi.nlm.nih.gov/cdd/
	Simple Modular Architecture Research Tool (SMART)	http://smart.embl-heidelberg.de/
	InterPro	http://www.ebi.ac.uk/interpro/
Identification of post translational modification sites	PROSITE	http://prosite.expasy.org/
Sequence similarity search	BLASTp (NCBI)	http://blast.ncbi.nlm.nih.gov/

Table 1. Databases and web-based programs used in the analysis of nuclear transport of *T. brucei*.

We utilized a personal computer equipped with AMD Turion 64x2 dual-core processor, memory size of 32 gigabytes and NVIDIA graphics card to perform the analyses. Our *in silico* work is summarized in Figure 4.

The nuclear transport refers to a process of entry and exit of large molecules from the cell nucleus. To identify *T. brucei* proteins of nuclear transport, the protein sequences of other various eukaryotic organisms were retrieved in FASTA format from National Centre for Biotechnology Information (NCBI) server and Universal Protein Knowledgebase/SwissProt (UniProtKB/ SwissProt) database based on biological processes and protein name search. The number of hits obtained for the query was recorded after manual inspection. The retrieved protein sequences were clustered into groups with more than 30% similarity using

BLASTClust [10] to reduce non-redundant protein sequences. The non-redundant data set was subjected to BLASTp [11] analyses against an integrated genomic and functional genomic database for eukaryotic pathogens of the family Trypanosomatidae, TriTrypDB. The analysis was using cutoff point with E-value of less than 1e-06 and score of more than 100. Hits that pointed to the same location or overlapped location were removed manually. The identified protein sequences then were then retrieved from the TriTrypDB.

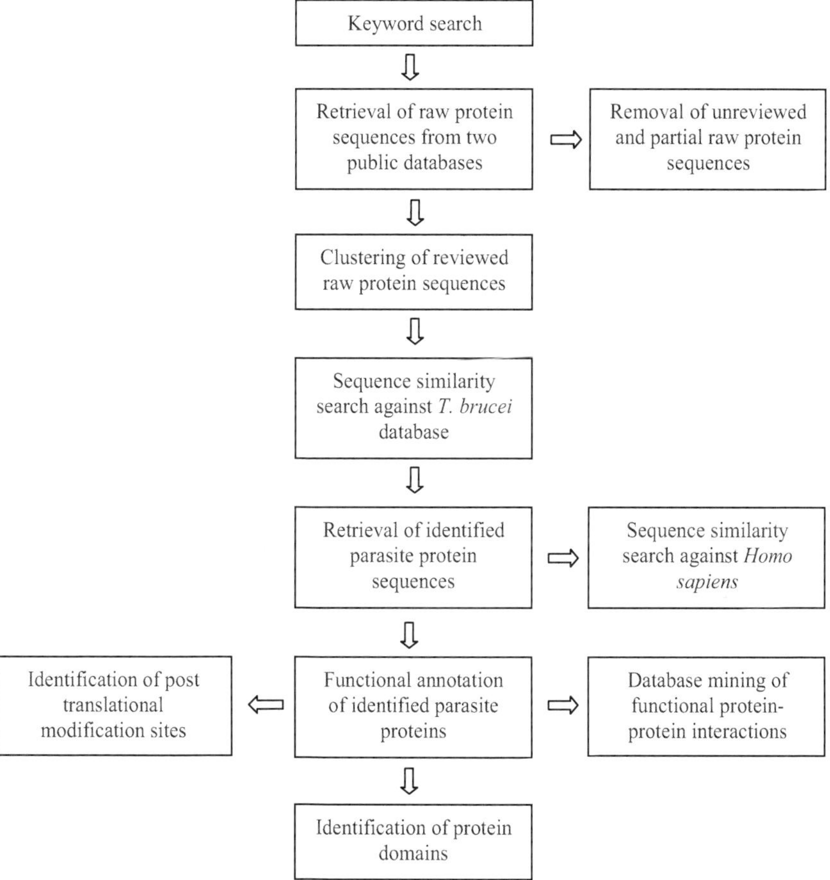

Figure 4. *In silico* analysis workflow.

A portion of protein that can evolve, function, and exist independently is called protein domain. It is a compact three dimensional structure, stable and distribution of polar and non-polar side chains contribute to its folding process. To determine the functional protein domains, all identified protein sequences of *T. brucei* from TriTrypDB were subjected to more than 2000 proteins are shuttled between the nucleus and the cytoplasm in yeast [19]. From our result, with the identification of Karyopherin and Nucleoporin proteins in *T. brucei*, we expect that the parasite employs the typical components for the nuclear transport machinery.

Subject sequences	E-value	Score	Functional protein domains
Tb927.3.1120	1.70E-72	718	Ran GTPase, GTP-binding domain
Tb09.211.4360	5.50E-33	348	Karyopherin Importin Beta, Armadillo repeat
Tb11.01.5940	9.30E-149	1391	Exportin-1 C terminal, Importin Beta N terminal domain
Tb11.02.0870	3.20E-16	187	Ran binding domain
Tb927.2.2240	2.40E-15	190	Exportin-like protein
Tb927.6.2640	9.10E-83	815	Karyopherin Importin Beta, Armadillo repeat
Tb927.6.4740	1.10E-75	748	CAS/CSE domain, Importin Beta N terminal domain
Tb927.7.1190	6.90E-20	172	RCC1 repeat
Tb11.03.0140	5.80E-09	107	NUP C terminal domain
Tb927.10.8170	2.10E-28	315	NUP C terminal domain
Tb927.8.3370	2.50E-48	281	Ran-binding protein Mog1p
Tb11.01.7010	8.20E-42	464	Armadillo repeat, Karyopherin Importin Beta
Tb11.02.1720	2.60E-26	276	Armadillo-like helical
Tb11.01.8030	1.70E-18	218	HEAT repeat, Armadillo repeat, Importin Beta N terminal domain
Tb11.01.7200	7.10E-07	137	Nsp1-like
Tb927.7.6320	1.20E-11	136	RCC1 repeat
Tb927.3.4600	3.70E-08	149	Armadillo-like helical
Tb09.160.2360	1.40E-36	379	WD40 repeat
Tb927.6.3870	8.50E-14	164	RNA recognition motif
Tb927.7.5760	1.30E-08	115	Nuclear transport factor 2 domain
Tb10.70.4720	4.60E-77	761	Importin Beta N terminal domain, Karyopherin domain
Tb927.8.4280	2.90E-08	112	Nuclear transport factor 2 domain

Key:

GTP	Guanine triphosphate
CAS	Cell apoptosis susceptibility
CSE	Chromosome seggregation
RCC1	Regulator of chromosome condensation 1
NUP	Nucleoporin
HEAT	Huntingtin, elongation factor 3 (EF3), protein phosphatase 2A (PP2A), and the yeast PI3-kinase TOR1
WD	Trp-Asp (W-D) dipeptide
RNA	Ribonucleic acid

Table 3. Identified and characterized *T. brucei* proteins of nuclear transport. Protein domain identification involved CDD, SMART, InterPro and PROSITE programs.

Protein domain	Accession	Description
Ran GTPase	SM00176	Ran is involved in the active transport of proteins through nuclear pores.
Ran binding domain	PDOC50196	This domain binds RanGTP and increases the rate of RanGAP1-induced GTP hydrolysis.
Armadillo	IPR000225	The Armadillo (Arm) repeat is an approximately 40 amino acid long tandemly repeated sequence motif first identified in the *Drosophila melanogaster* segment polarity gene armadillo involved in signal transduction through wingless. Animal Arm-repeat proteins function in various processes, including intracellular signalling and cytoskeletal regulation, and include such proteins as beta-catenin, the junctional plaque protein plakoglobin, the adenomatous polyposis coli (APC), tumour suppressor protein, and the nuclear transport factor importin-alpha, amongst others
Importin beta	IPR001494	Members of the Importin-beta (Karyopherin-beta) family can bind and transport cargo by themselves, or can form heterodimers with importin-alpha. As part of a heterodimer, Importin-beta mediates interactions with the pore complex, while Importin-alpha acts as an adaptor protein to bind the nuclear localisation signal (NLS) on the cargo through the classical NLS import of proteins.
HEAT	IPR000357	Arrays of Huntingtin, elongation factor 3 (EF3), protein phosphatase 2A (PP2A), and the yeast PI3-kinase TOR1 (HEAT) repeats consists of 3 to 36 units forming a rod-like helical structure and appear to function as protein-protein interaction surfaces. It has been noted that many HEAT repeat-containing proteins are involved in intracellular transport processes.
Exportin 1-like protein	pfam08389	The sequences featured in this family are similar to a region close to the N-terminus of yeast exportin 1 (Xpo1, Crm1). This region is found just C-terminal to an importin-beta N-terminal domain (pfam03810) in many members of this family. Exportin 1 is a nuclear export receptor that interacts with leucine-rich nuclear export signal (NES) sequences, and Ran-GTP, and is involved in translocation of proteins out of the nucleus.

Protein domain	Accession	Description
CAS/CSE	IPR005043	In the nucleus, cell apoptosis susceptibility (CAS) acts as a nuclear transport factor in the importin pathway. The Importin pathway mediates the nuclear transport of several proteins that are necessary for mitosis and further progression. CAS is therefore thought to affect the cell cycle through its effect on the nuclear transport of these proteins
WD40	IPR001680	WD-repeat proteins are a large family found in all eukaryotes and are implicated in a variety of functions ranging from signal transduction and transcription regulation to cell cycle control and apoptosis. Repeated WD40 motifs act as a site for protein-protein interaction, and proteins containing WD40 repeats are known to serve as platforms for the assembly of protein complexes or mediators of transient interplay among other proteins.
RCC1	PDOC00544	The regulator of chromosome condensation (RCC1) is a eukaryotic protein which binds to chromatin and interacts with ran, a nuclear GTP-binding protein (see <PDOC00859>), to promote the loss of bound GDP and the uptake of fresh GTP, thus acting as a guanine-nucleotide dissociation stimulator (GDS)
NUP C-terminal	PDOC51434	Communication between the nucleus and cytoplams of an eukaryotic cell is mediated by the nuclear pore complexes (NPCs), which act as selective molecular gateways. Through these gateways, ribonucleic acids (RNAs) and proteins are exported into the nucleus. Each NPC consists of ~30 distinct proteins termed Nucleoporins, each present in at least eight copies, reflecting the octagonal symmetry of the complex.
NSP 1	IPR007758	The NSP1-like protein appears to be an essential component of the nuclear pore complex, for example preribosome nuclear export requires the Nup82p-Nup159p-Nsp1p complex.
NTF 2	IPR002075	Nuclear transport factor 2 (NTF2) is a homodimer which stimulates efficient nuclear import of a cargo protein. NTF2 binds to both RanGDP and FxFG repeat-containing Nucleoporins.

Table 4. Summary of protein domains

3.2. Regulatory aspect of the parasite nuclear transport

Table 5 shows the presence of phosphorylation and glycosylation sites in the parasite proteins. The phosphorylation sites were found to be present in all parasite proteins. It was predicted that the parasite proteins could be phosphorylated at Serine, Threonine and Tyrosine amino residues. However, the O-glycosylation sites were not present in three parasite proteins, namely Tb11.02.0870, Tb927.8.3370 and Tb927.7.5760.

Subject sequences	Phosphorylation site	Glycosylation site
Tb927.3.1120	+	+
Tb09.211.4360	+	+
Tb11.01.5940	+	+
Tb11.02.0870	+	-
Tb927.2.2240	+	+
Tb927.6.2640	+	+
Tb927.6.4740	+	+
Tb927.7.1190	+	+
Tb11.03.0140	+	+
Tb927.10.8170	+	+
Tb927.8.3370	+	-
Tb11.01.7010	+	+
Tb11.02.1720	+	+
Tb11.01.8030	+	+
Tb11.01.7200	+	+
Tb927.7.6320	+	+
Tb927.3.4600	+	+
Tb09.160.2360	+	+
Tb927.6.3870	+	+
Tb927.7.5760	+	-
Tb10.70.4720	+	+
Tb09.211.2550	+	+
Tb927.8.4280	+	+

Key:

(+) indicates presence
(-) indicates absence

Table 5. Phosphorylation and O-glycosylation sites in the *T. brucei* proteins. Identification of these functional sites involved ScanProsite programme.

Most of the parasite proteins were predicted to be involved in O-linked glycosylation. In eukaryotes, the O-linked glycosylation takes place in the Golgi apparatus. It also occurs in archaea and bacteria. Phosphorylation was reported to be crucial in the regulation of protein-protein interactions of the NADPH oxidase in the phagocytic cells [20]. The phosporylation-based signaling in *T. brucei* has been reported by reference [21]. Thus we procedure to partially purify *T. cruzi* ubiquitin was performed. Following this preparation, ELISA and Western blots were carried out to show that chagasic sera recognise *T. cruzi* but not human or Leishmania ubiquitin indicating a species-specific response. Thus, it is probable that the *T. brucei* proteins could also be distinguished from human counterparts at primary sequence level by using the immunodetection method.

4. General discussions

4.1. Transport of cargoes

In RanGTPase system, Ran-binding protein 1 (RanBP1) which is cytoplasmic localized binds RanGTP and eases the RanGAP-dependent conversion of RanGTP to RanGDP [29]. This indicates that RanBP1 catalyses the cytoplasmic disassembly of RanGTP–transport receptor complexes. These complexes are kinetically so stable that RanGAP alone fails to trigger GTP hydrolysis [30-32]. RanBP2 [33] is a major constituent of the cytoplasmic filaments of NPCs and exhibits similar functions as RanBP1. It has four RanBP1 homology domains and forms a stable complex with sumoylated RanGAP [34,35], in order to dismantle the RanGTP–transport receptor complexes that exit the nucleus. Importin- and exportin-mediated transport cycles can accumulate cargoes against gradients of chemical activity, which is an energy-dependent process. The RanGTPase system hydrolyses one GTP molecule per transport cycle, and a number of evidences suggest that this contains the sole input of metabolic energy [36-39]. We have successfully identified all the required key components in the *T. brucei* nuclear transport. Whether their functionalities *in vivo* are consensus with the known ones still remains to be further investigated.

4.2. Relationship between signaling pathways and nuclear transport

Many aspects of cell physiology are greatly dependent on the signaling pathways. This includes members of the mitogen activated protein (MAP) kinase family as well as phosphatidyl inositol 3 (PI3) and adenosine monophosphate (AMP) kinases which are crucial in controlling the cell growth, proliferation, apoptosis and the response to stress. By activating the signaling pathway through multiple kinase cascades, various stressors are able to regulate the nuclear transport. For example, oxidative and heat stress activate both MAP kinase kinase (MEK)-extracellular signal regulated kinase 1/2 (ERK1/2) and PI3 kinase-Akt pathways [40]. Based on these observations and the fact that many of the transport components are modified post-translationally, it was sensible to investigate whether these modifications are regulated by stress. A study reported by [41] showed that oxidant treatment induced phosphorylation and/or GlcNAc modification of soluble transport factors and nucleoporins. Interestingly, changes in transport factor modifications are not limited to stress conditions, as modifying ERK or PI3 kinase activities in unstressed cells also affect the transport factors. This is exemplified by the regulation of RanBP3 through ERK1/2-ribosomal S6 kinase (RSK) signaling, a regulatory link which ultimately controls the Ran concentration gradient. Furthermore, phosphorylation of Nup50 which is dependent on ERK, reduces its association with importin-β1 and transportin *in vitro*, and ERK2 is responsible to the oxidant-induced collapse of the Ran gradient [42]. It remains unknown how much modulating individual transport factors contributes to the overall regulation of nuclear trafficking. However, it is noteworthy that the kinase inhibitor PD98059, which targets ERK1/2 and ERK5, significantly increases classical nuclear import, both under normal and stress conditions. Taken together, these results highlight a critical role of ERK activity in nuclear transport, with ERK kinases targeting both soluble factors and nucleoporins [41]. Thus, there is an urgent need to investigate the possible connection between upstream signaling apparatus with nuclear transport components in *T. brucei*.

4.3. *In silico* approach for drug target discovery

We have provided interpretation of heterologous data sets for nuclear transport system of *T. brucei* from various resources. With the availability of protein databases and computer-aided softwares, we are able to explain various functional interactions between identified parasite proteins and how these functional interactions give rise to functionality and behavior of the parasite nuclear transport. This would partially facilitate the exhausted effort to obtain system-level understanding of *T. brucei* pathogenesis. Our *in silico* approach has the potential to speed up the rate of drug target discovery while reducing the need for expensive lab work and clinical trials. The conventional approaches *in vivo* and *in vitro* have high tendencies to produce inefficient results when investigating complex large scale data such as proteins associated with nuclear shuttling of macromolecules across the nuclear envelope. Therefore, the systematic *in silico* approach from this study provides a tremendous opportunity of cost effective drug target discovery for the pharmaceutical industry.

4.4. Experimental validation of *in silico* data

Experimental techniques such as yeast two-hybrid assay and affinity purification combined with mass spectrometry are useful to investigate the possible protein-protein interaction. However, they have their limitations in detecting certain types of interactions. They also have technical problems to scale-up for high-throughput analysis. In conjunction with this, *in silico* approach may solve those problems in inferring the protein function. The scope of experimental data can be expanded to increase the confidence of certain interacting protein pairs with the availability of databases containing *in silico* data such as protein domain and 3D structure. The databases integrate information from various resources such as computational prediction methods and public text collections. Since *in silico* and experimental approaches are complementary to each other, the combination of these different approaches is very useful to obtain a more accurate picture of *T. brucei* nuclear transport.

4.5. Our further direction

In silico approach offers various advantages over *in vivo* and *in vitro* approaches such as non-use of animals, low costs, and reduced execution time. This approach allows identification of proteins of interest from a particular biological study. From a protein function standpoint, transfer of annotation from known proteins to a novel target is currently the only practical way to convert vast quantities of raw sequence data into meaningful information. Many bioinformatics tools now provide more sophisticated methods to transfer functional annotation, integrating sequence, family profile and structural search methodology. Thus, in addition to data mining for protein-protein interaction, further *in silico* approach should also consider structural alignment, molecular docking and pathway modeling in order to obtain a comprehensive and more reliable insight into protein-protein interaction of *T. brucei* nuclear transport.

5. Conclusion

The availability of protein databases and computer-aided softwares to identify probable components of cellular mechanisms has become a new trend in the present scientific era. We demonstrate here a computational analysis of nuclear transport in *T. brucei* as an initial step and proof of concept for further investigation. Our approach successfully identified 22 *T. brucei* proteins essential for nuclear transport. All those parasite proteins were found to contain relevant functional domains that drive the translocation of macromolecules in the parasite. The phosphorylation and O-glycosylation sites were also detected in all identified parasite proteins. This has given us an insight into the regulatory aspect of parasite nuclear transport. The database mining of protein interaction has shown that nine out of 22 parasite proteins possess relevant functional interactions for nuclear transport activities. However, more functional interactions from nuclear transport constituents of *T. brucei* are required to elucidate the exact mechanism. The homology between the parasite proteins and human counterparts was shown by BLASTp analyses. Whether there are structural differences between them remain unknown.

The nuclear transport in *T. brucei* has been characterized by using the *in silico* approach. The predicted functionalities and regulatory aspects of parasite nuclear transport constituents were in agreement with the previous reports. Moreover, the protein interaction data derived from the public database has made the participation of parasite proteins in the mechanism more convincing. Thus, we have laid a path for understanding the nuclear transport machinery in *T. brucei*. The development of drugs that target as well as alter nuclear import and export will undoubtedly become beneficial in controlling Trypanosomiasis in future. Drugs that have a direct effect on a single protein must be able to localize to the same site as the protein and interact with one or more of its domains. Alternatively, a drug that effectively blocks the target protein from reaching its proper organelle can also inhibit the protein's function.

Author details

Mohd Fakharul Zaman Raja Yahya[*] and Umi Marshida Abdul Hamid
School of Biology, Faculty of Applied Sciences, MARA University of Technology Shah Alam, Shah Alam Selangor, Malaysia

6. References

[1] Barrett M P, Burchmore R J and Stich A (2003). The trypanosomiases. Lancet 362 (9394): 1469–80.

[2] Miles M (2003). American trypanosomiasis (Chagas disease). GC Cook, A Zumla (Eds.), Manson's tropical disease (21st edn.), Elsevier Science, London, pp. 1325–1337.

[3] Hecker H and Bohringer S (1977). Morphometric analysis of the life cycle of *Trypanosoma brucei*. Ann. Soc. belge Med. trop., vol. 57 (4-5), pp. 465-470.

[4] Ogbadoyi E, Ersfeld K, Robinson D, Sherwin T, Gull K (2000). Architecture of the *Trypanosoma brucei* nucleus during interphase and mitosis. Chromosoma 108 (8): 501–13.

[5] Acosta-Serrano A, Vassella E, Liniger M, Renggli C K, Brun R, Roditi I and Englund P T. (2001). The surface coat of procyclic *Trypanosoma brucei*: Programmed expression and proteolytic cleavage of procyclin in the tsetse fly. PNAS, vol. 98(4), pp. 1513-1518, 2001.

[6] Gorlich D and Mattaj I W (1996). Nucleocytoplasmic transport. Science vol. 271 (5255), pp. 1513-1518.

[7] Mattaj I W and Englmeier L (1998). Nucleocytoplasmic transport: The Soluble Phase. Annual Review of Biochemistry, vol. 67, pp. 265-306.

[8] Fried, H., and Kutay, U. (2003). Nucleocytoplasmic transport:taking an inventory. Cell Mol Life Sci 60, 1659–1688.

[9] Röhrig U F, Awad L, Grosdidier A, Larrieu, P, Stroobant V, Colau D, Cerundolo V and Andrew J. G. (2010). Rational Design of Indoleamine 2,3-Dioxygenase Inhibitors. Journal of Medicinal Chemistry 53 (3): 1172–89.

[10] Hayes, C N (2008). varDB: a pathogen-specific sequence database of protein families involved in antigenic variation, Bioinformatics.

[11] Altschul S F, Gish W, Miller W, Myers E W, and Lipman D J. (1990). Basic local alignment search tool. J. Mol. Biol, vol. 215, pp. 403-410.

[12] Marchler-Bauer A, Anderson J B, Cherukuri P F, DeWeese-Scott C, Geer L Y, Gwadz M, He S, Hurwitz D I, Jackson J D, Ke Z, Lanczycki C J, Liebert C A, Liu, Lu C F, Marchler G H, Mullokandov M, Shoemaker B A, Simonyan V, Song J S, Thiessen P A, Yamashita R A, J. J. Yin, D. Zhang, and S. H. Bryant (2005). CDD: a Conserved Domain Database for protein classification. Nucleic Acid Research , vol. 33, pp. 192-196.

---

[*] Corresponding Author

[13] Letunic I, Doerks T and Bork P (2009). SMART 6: recent updates and new developments. Nucleic Acid Research. vol. 37, pp. 229-232.

[14] Hunter S, Apweiler R, Attwood T K, Bairoch A, Bateman A, Binns D, Bork P, Das U, Daugherty L, Duquenne L, Finn R D, Gough J, Haft D, Hulo N, Kahn D, Kelly E, Laugraud A, Letunic I, Lonsdale D, Lopez R, Madera M, Maslen J, McAnulla C, J. McDowall, Mistry J, Mitchell J A, Mulder N, Natale D, Orengo C, Quinn A F, Selengut J D, Sigrist C J, Thimma M, Thomas P D, Valentin F, Wilson D, Wu C H, and Yeats C (2009). InterPro: the integrative protein signature database. Nucleic Acids Research, vol. 37, pp. 211-215.

[15] Hulo N, Bairoch A, Bulliard V, Cerutti L, De Castro E, Langendijk-Genevaux D S, Pagni M, and Sigrist C J A (2006). The PROSITE database. Nucleic Acid Research, vol. 34, pp. 227-230.

[16] Jensen L J, Kuhn M, Stark M, Chaffron S, Creevey C, Muller J, Doerks T, Julien P, Roth A, Simonovic M, Bork P, and Von Mering C (2009). STRING 8--a global view on proteins and their functional interactions in 630 organisms. Nucleic Acid Research. Vol. 7 pp 412-416.

[17] Schmid M (1998). Novel approaches to the discovery of antimicrobial agents. Curr. Opin. Chem. Biol. vol. 2, pp. 529-534.

[18] Frankel M B and Knoll L J (2009). The Ins and Outs of Nuclear Trafficking: Unusual Aspects in Apicomplexan Parasites. DNA and Cell Biology vol. 28, pp. 277-284.

[19] Macara, I G (2001). Transport into and out of the nucleus. Microbiol Mol Biol Rev 65, 570–594.

[20] Babior B M (1999). NADPH oxidase: an update. Blood 93 (5): 1464–76.

[21] Nett I R E, D. Martin D M A, Miranda-Saavedra D, Lamont D, Barber J D, and Mahlert A (2009). The phosphoproteome of bloodstream form *Trypanosoma brucei*, causative agent of African sleeping sickness. Molecular & Cellular Proteomics, vol. 8 pp. 1527-1538.

[22] Miller, M. W., Caracciolo, M. R., Berlin, W. K., and Hanover, J. A. (1999). Phosphorylation and Glycosylation of Nucleoporins. Archives of Biochemistry and Biophysics, 367(1), 51-60.

[23] Gasiorowski, J. Z. and Dean, D. A. (2003). Mechanisms of nuclear transport and interventions. Advanced Drug Delivery Reviews, 55, 703-716.

[24] Schuldt, A. (2012). Post-translational modification: A monoubiquitylation pore anchor. Nature Reviews Molecular Cell Biology, 13, 66.

[25] Radu A, Blobel G and Moore M S (1995). Identification of a protein complex that is required for nuclear protein import and mediates docking of import substrate to distinct nucleoporins, Proc. Natl. Acad. Sci. USA 92: 1769– 1773.

[26] Gasiorowski, J. Z. and Dean, D. A. (2003). Mechanisms of nuclear transport and interventions. Advanced Drug Delivery Reviews, 55, 703-716.

[27] Damelin M, S. P. (2000). Mapping interactions between nuclear transport factors in living cells reveals pathways through the nuclear pore complex. Mol Cell., 5(1), 133-40.

[28] Télles S, Abate T and Slezynger T C (1999). *Trypanosoma cruzi* and human ubiquitin are immunologically distinct proteins despite only three amino acid difference in their primary sequence. FEMS Immunol Med Microbio, 24(2), 123-30.

[29] Bischoff F R, Krebber H, Smirnova E, Dong W H, and Ponstingl H. (1995). Coactivation of RanGTPase and inhibition of GTP dissociation by Ran GTP binding protein RanBP1. EMBO J, vol. 14, pp. 705–715.

[30] Bischoff F R and Görlich D. (1997). RanBP1 is crucial for the release of RanGTP from importin β -related nuclear transport factors. FEBS Lett, vol. 419, pp. 249–254.

[31] Floer M, Blobel G and M. Rexach M (1997). Disassembly of RanGTP–karyopherin β complex, an intermediate in nuclear protein import. J Biol Chem, vol. 272, pp. 19538–19546.

[32] Lounsbury K M and Macara I G (1997). Ran-binding protein 1 (RanBP1) forms a ternary complex with Ran and karyopherin β and reduces Ran GTPase-activating protein (RanGAP) inhibition by karyopherin β. J Biol Chem, vol. 272, pp. 551–555.

[33] Yokoyama N (1995). A giant nucleopore protein that binds Ran/TC4. Nature, vol. 376, pp. 184–188.

[34] Mahajan R, Delphin C, Guan T, Gerace L and Melchior F (1997). A small ubiquitin-related polypeptide involved in targeting RanGAP1 to nuclear pore complex protein RanBP2. Cell, vol. 88, pp. 97–107.

[35] Matunis M J, Coutavas E, and Blobel G (1996). A novel ubiquitin-like modification modulates the partitioning of the Ran-GTPase-activating protein RanGAP1 between the cytosol and the nuclear pore complex. J Cell Biol, vol. 135, pp. 1457–1470.

[36] Englmeier L, Olivo J C and Mattaj I W (1999). Receptor-mediated substrate translocation through the nuclear pore complex without nucleotide triphosphate hydrolysis. Curr Biol, vol. 9, pp. 30–41.

[37] Kose S, Imamoto N, Tachibana T, Shimamoto T, and Yoneda Y (1997). Ran-unassisted nuclear migration of a 97 kD component of nuclear pore- targeting complex. J Cell Biol, vol. 139, pp. 841–849.

[38] Ribbeck K, Kutay U, Paraskeva E and Görlich D (1999). The translocation of transportin–cargo complexes through nuclear pores is independent of both Ran and energy. Curr Biol, vol. 9, pp. 47–50.

[39] Weis K, Dingwall C, and Lamond A I (1996). Characterization of the nuclear protein import mechanism using Ran mutants with altered nucleotide binding specificities. EMBO J, vol. 15, pp. 7120–7128.

[40] Kodiha M, Banski P and Stochaj U (2009). Interplay between MEK and PI3 kinase signaling regulates the subcellular localization of protein kinases ERK1/2 and Akt upon oxidative stress. FEBS Lett, 583:1987-93.

both lentiviruses had a preferential integration close to Alu elements which correspond to SINEs. Within LINEs, differences among L1, L2 and L3 were recorded. The other class of repetitive elements like LTR, simple repeats and low complexity represented a minor proportion of the integration associated chromatin (figure 3).

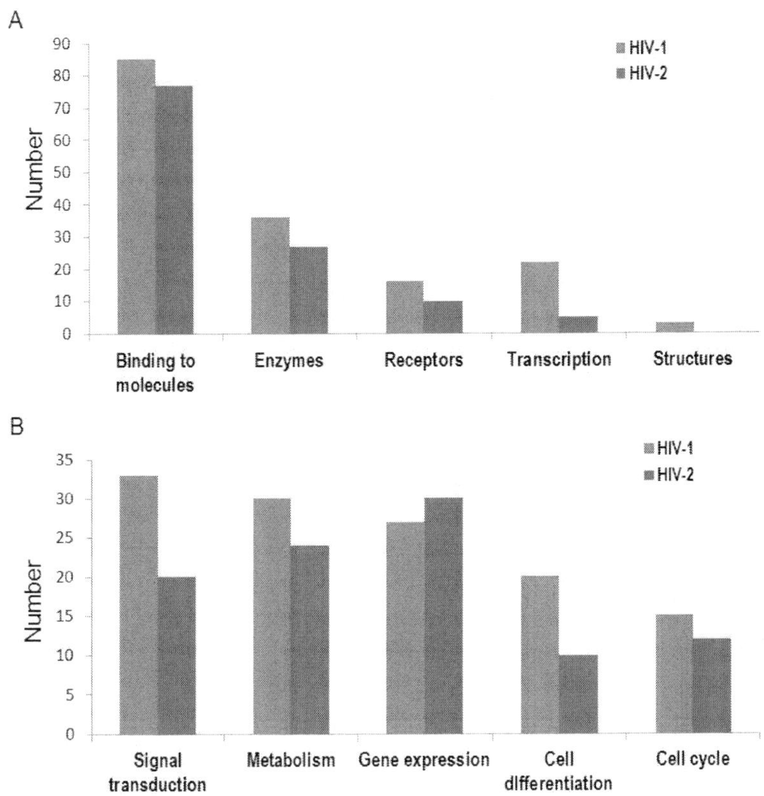

Figure 2. Functional characterization of the coding protein genes located in genome regions around 100 kb of human lentivirus. (a) Molecular function by GO of genes associated with HIV-1 and HIV-2 integration. (b) Biological process by GO of genes associated with HIV-1 and HIV-2 integration. Blue blocks correspond to HIV-1. Red blocks to HIV-2

5.3. Definition of the common genomic environment of integrations

As the integration do not follow a random model [23-25], some characteristics of the chromatin associated with regions with high level of provirus integration, support the hypothesis that a preferential integration is conditioned by structural and functional states of local chromatin; these states are defined by several genomic variables which were studied in this work, and together would define genomic environments.

The results of multiple-regression analysis conducted on the HIV-1 and HIV-2 data sets showed that there were differential distributions of CpG island, genes, and Alu elements that together conditioned a specific genomic environment per chromosome (R^2=0.91, p<0.05). Gene density was the independent variable contributed most in the prediction of the dependent variable (integrations) due to the highest regression coefficients (B= 0.83; p<0.05). The highest relative likelihood of hosting a lentiviral integration event in the human genome was registered in chromosome 17 (figure 4a). To test that integration events are favored by gene-rich regions in all chromosomes, a comparison between those variables was done indicating that a high gene density in chromatin regions determine a favorable environment for integration, even when the chromosome 17 is excluded   (Figure 4b). Because chromosome 17 registered the highest percentage of Lentiviral integration events, a detailed analysis of chromatin structure correlating several variables that give data about the cellular chromatin status was performed. In general the distal chromatin regions of p and q arms showed similarities in the distribution of methylation in CpG islands, methylation in several lysine residues of histone H3 (K4, K27 and K36) and variable levels of open chromatin and nucleosome occupancy (figure 5a and b).

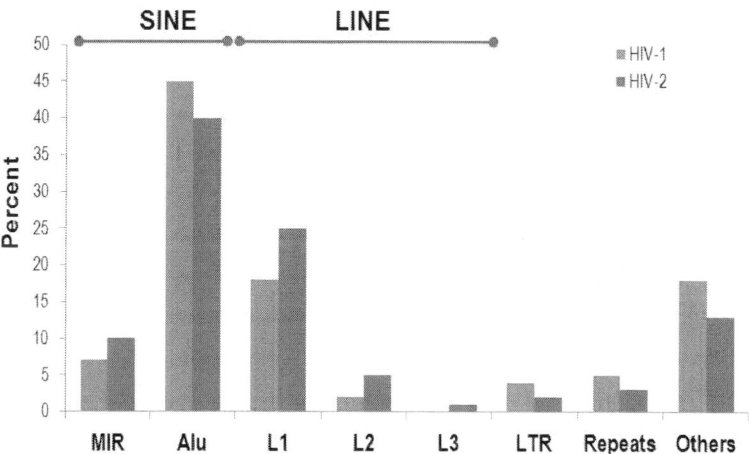

Figure 3. Frequencies of several repetitive elements associated with regions of 100 kb around the HIV-1 and HIV-2 proviruses. SINEs, short interspersed nuclear element; LINEs, long interspersed nuclear element.

Experimental studies have demonstrated that regulatory regions in general and promoters in particular, tend to be DNase sensitive and are target for integration of the majority of retroviruses [37, 38]. In 2006, the complete nucleotide sequence of chromosome 17 was published [39]. This chromosome is rich in protein coding genes, having the second highest gene density in the genome, (16.2 genes per Mb), with a relative excess of short interspersed elements (SINEs, 22.3%) and a deficit of long interspersed elements (LINEs, 14.4%). Likewise, this chromosome has high average CpG content (45.5%) and high euchromatin density [39] (figure 6). Our statistical analysis determined that chromosome 17 had the highest number of integrations, mainly concentrate towards the telomeres of both arms.

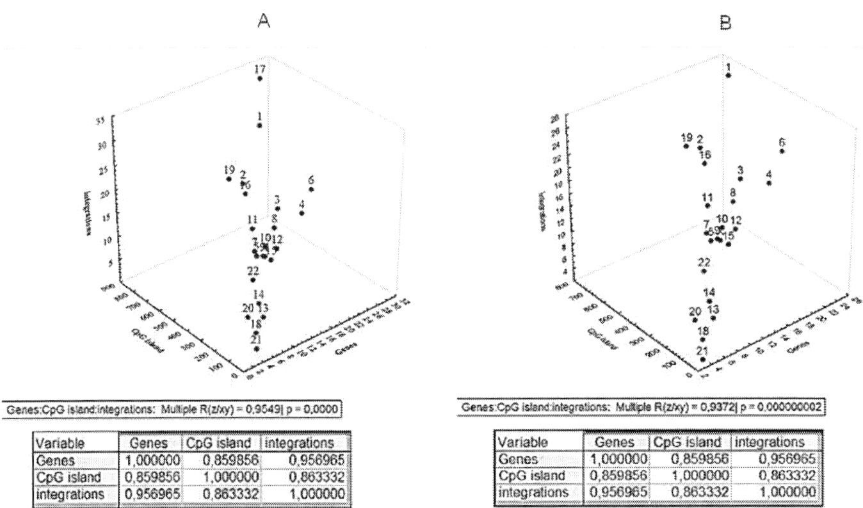

Genes:CpG island:integrations: Multiple R(z/xy) = 0,9549| p = 0,0000

Variable	Genes	CpG island	integrations
Genes	1,000000	0,859856	0,956965
CpG island	0,859856	1,000000	0,863332
integrations	0,956965	0,863332	1,000000

Genes:CpG island:integrations: Multiple R(z/xy) = 0,9372| p = 0,000000002

Variable	Genes	CpG island	integrations
Genes	1,000000	0,859856	0,956965
CpG island	0,859856	1,000000	0,863332
integrations	0,956965	0,863332	1,000000

Figure 4. Multiple-regression analysis among gene density, CpG island number, and frequency of HIV-1 and HIV-2 proviruses including every human chromosome. A high statistical correlation is observed mainly for chromosome 17. (a) Analysis including all human chromosomes. (b) The same analysis but excluding chromosome 17.

The most relevant relationship was related to the conformational state of chromatin including the nucleosomes occupancy, methylation of CpG Islands, DNase hypersensitive regions and transcriptionally active genes that are found in open-decondensed chromatin regions. These regions provide the environment for DNA regulatory processes such as DNA replication, repairs and transcription. Albanese et al. (2008) [40], found histone and IN acetylation may favor integration by tethering the virus to acetylated/decondensed regions of the chromatin. We concluded that the structural characteristics and the epigenetic modifications observed in those regions with high frequency of cDNA viral integrations would synergistically configure a local "genomic environment" that facilitates the target site selection during the retroviral integration.

5.4. Construction of HIV-1 gene/protein networks

Host-virus interactions is a complex level of systems information that permits a thorough understanding of how the virus exploits the host cell and uses the cellular machinery to integrate into host genome. Recently, the HIV-1 Human Protein Interaction Database (HHPID) registered 3959 interactions among 1452 human proteins and nineteen HIV proteins (fifteen of them structural and four intermediate proteins) [41].

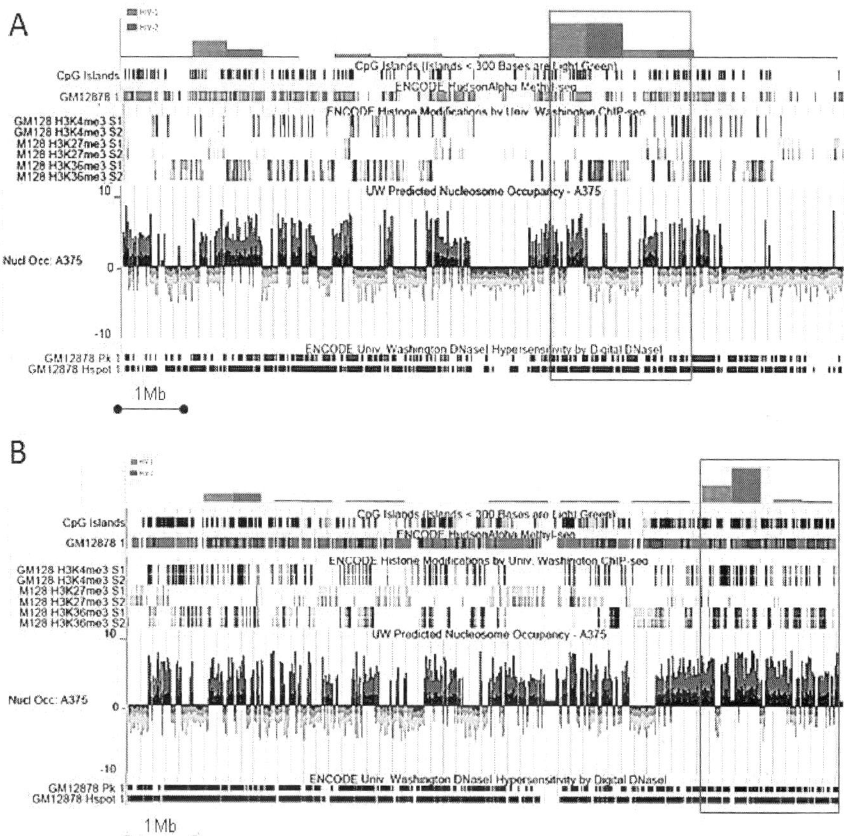

Figure 5. Flash image of the Genome Browser showing the distribution of several characteristics of the chromatin along 9.5 Mb of the p and q arm representing 25% of chromosome 17. (a) p arm, (b) q arm. The figure shows the GC percentage for each: 5pb (black), refseq Genes (black), CpG islands (several tones of gray), levels of open chromatin (ENCODE, Duke) in GM12878 cells with DNaseI and FAIRE (Formaldehyde Assisted Isolation of Regulatory Elements) (black), DNase1 hypersensitivity (ENCODE, University of Washington) in GM12878 cells (gray), pk (sites identified as signal peaks within FDR 0.5% hypersensitive zones), Hspots (zones identified using the HotSpot algorithm), and predicted nucleosome occupancy in A375 cells (black peaks).

Previous studies have identified most of human cell pathways been disturbed by at least one interaction with an HIV-1 protein during the virus life cycle [42-44]. Those interactions are of two types: either direct, via host cell protein-viral proteins or indirect, such as regulatory interactions that alter expression of human genes [45, 46]; the signaling network cc-cytokine is both disrupted and exploited by HIV at various stages of infection. 22 candidates human class E proteins were connected into coherent network by 43 different protein-protein interactions, in which AIP1 play a key role in linking complexes that act early (TSG101/ESCRT-I) and late (CHMP4/ESCRT-III) in the HIV infection pathways [47, 49]. Monocyte/macrophage infection is characterized by a viral dynamic substantially different from that of T lymphocytes. In fact, *in vivo* HIV infection of activated CD4-T lymphocytes accounts for the majority of the daily production of virus particles. However, a large number of lymphocytes are in a resting state, thus unable to sustain a complete and productive virus life cycle, and contribute only minimally to the daily virus production [50-52]. Because of the limited HIV-induced cytopathic effect and of their ability to accumulate high levels of HIV particles in intracellular compartments, HIV-infected macrophages serve as a potentially important reservoir, and as "Trojan horses" exploited by the virus to favor its dissemination in different tissues. [53, 54].

Cytoscape v.2.63 [55] was also used to construct a gene expression network from two kinds of files: The first one from gene expression profiles as a text file (.pvals) that were imported of expression data microarray experiments (GEO profiles, NCBI). The second, as data annotation in text files (.sif) that corresponds to each one gene-gene interactions (online databases). In the first one, gene expression values were collected from the microarray data series GSE19236 composed by two Agilent platforms (**GPL6480** and **GPL6848**) with 48 samples of monocytes to macrophages, macrophages and dendritic cells. These are available from the National Center for Biotechnology Information (NCBI) Gene Expression Omnibus (GEO) repository (accession number GEO: GSE19236) and for our analysis, we selected all macrophages expression samples (GSM476720, GSM476721, GSM476722, GSM476723, GSM476724, GSM476725). To identify which genes were significant among samples in microarrays; considering a p-value< 0.001 as significant, an ANOVA test was calculated. Additionally, a Hierarchical clustering analysis of the samples using Euclidean Distance Method and mean linking were performed. MultiExperiment Viewer v4.1 [56] was applied to make the corresponding statistical analyses. Using data from BOND (Biomolecular Network Data Bank, http://bond.unleashedinformatics.com/Action), BioGird (Biological General Repository for Interaction Datasets, http://thebiogrid.org/), KEEG (Kyoto Encyclopedia of Genes and Genomes, http://www.genome.jp/kegg/), available online, a new file with the interaction data of 28 genes located close to integration sites was constructed.

Cytoscape v2.6 was used for visualizing and analyzing the genetic interaction networks among 28 human macrophages genes and their interactions. BiNGO v2.6 plugin (Biological Networks Gene Ontology tool) was used to determine which Gene Ontology (GO) terms are significantly overrepresented in a set of genes. A hypergeometric test was applied to determine which categories were significantly represented (p-value< 0.01); significant value was adjusted for multiple hypotheses testing using the Bonferroni Family-wise error rate correction [57]. The network topology parameters were calculated using Network Analyzer plug-in, which includes network diameter, the number of connected pairs of nodes and average number of neighbors; it also analyses node degrees, shortest paths, clustering coefficients, and topological coefficients (Max Planck Institute Informatik).

To identify active sub-networks as highly connected regions of the main network we used j ActiveModules plug-in that grouped genes according with significant p-values of gene expression over particular subsets of samples. The result shows active modules, listed according to the number of nodes, and an associated Z-score. An active module with Z-scores greater than 3.0 indicated significant response upon the conditions of the experiment. We kept the standard default values, as being the most effective for initial analyses (58).

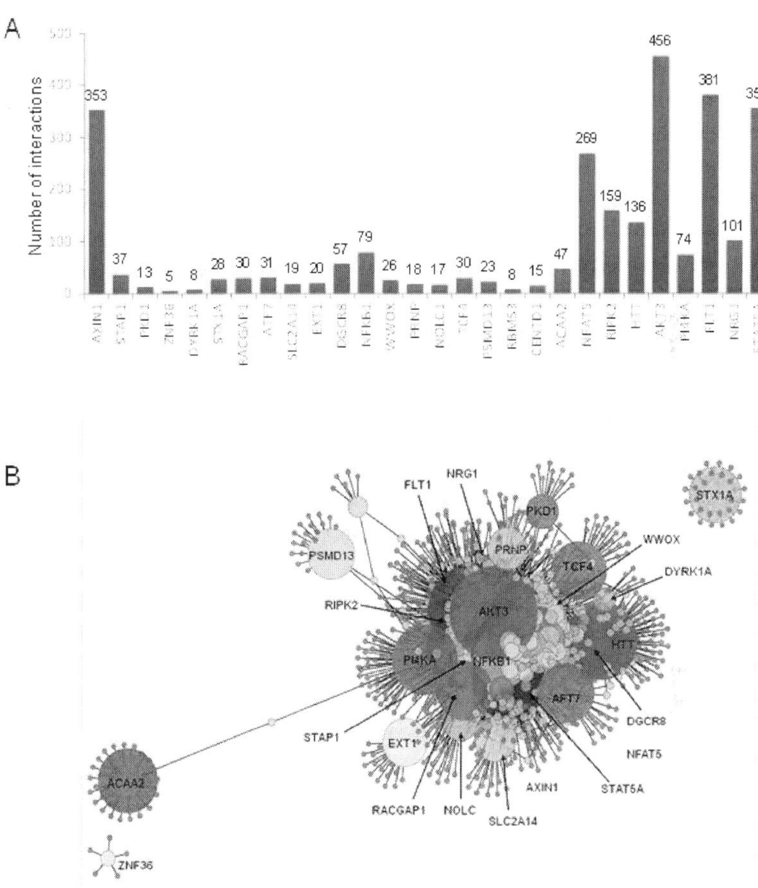

Figure 6. (A). Interaction values of 28 genes of human macrophages interrupted by HIV-1 cDNA integration. (B). Gene expression network in non HIV-1 infected macrophage. Visualization of gene network composed by 28 genes located close to regions with high frequency of HIV-1 provirus in human macrophages. These genes interact with 1202 genes through 2770 interactions. The network was constructed using Cytoscape. Each node corresponds to a gene and edges represent interactions among genes. The color gradient represents the expression values

Eleven thousand and seven hundred and thirteen (11,713) significant genes of 41,000 probes were clustered in two significant different groups of cells; one of them included only could be further classified into two major groups; cell function regulation and signaling of biological process. In contrast HIV-1 infected macrophage gene network was enriched with 10 significant functional categories of a total of 40. The significantly overrepresented categories indicated that this emergent new gene network was composed by genes involved in metabolic process and DNA repair process.

In this study we simulated at systemic level, the alterations of cellular pathways when HIV provirus integrates into genes by turning them off and produce dysregulation of several local signaling pathways. One of the target gene associated with HIV-1 integration was AKT3, also called PKB, which is a serine/threonine protein kinase family member. It is involved in a wide range of biological processes including cell proliferation, differentiation, apoptosis, stimulating cell growth, and regulating other biological responses (59, 60). Also, it have been identified playing important roles of regulation in the G2/M transition of the cell cycle.

Normal Network	p-value [b]	HIV-1 infected Network	p-value [b]
Signal transduction [GO-ID: 7165]	3,90E-123	Biopolymer biosynthetic process [GO-ID: 43284]	6,27E-05
Cell communication [GO-ID: 7154]	4,63E-120	Metabolic process [GO-ID: 8152]	8,00E-05
Positive regulation of cellular process [GO-ID: 51242]	1,39E-95	Macromolecule biosynthetic process [GO-ID: 9059]	1,40E-03
Positive regulation of biological process [GO-ID: 48518]	1,93E-92	Biosynthetic process [GO-ID: 9058]	2,43E-03
Biological regulation [GO-ID: 65007]	1,45E-88	alcohol metabolic process [GO-ID: 6066]	3,19E-03
Intracellular signaling cascade [GO-ID: 7242]	1,40E-86	Maintenance of fidelity during DNA-dependent DNA replication [GO-ID: 45005]	6,24E-03
Regulation of cellular process [GO-ID: 51244]	2,29E-85	Mismatch repair [GO-ID: 6298]	6,24E-03
Regulation of biological process [GO-ID: 50791]	2,65E-84	Furaldehyde metabolic process [GO-ID: 33859]	6,68E-03
Phosphate metabolic process [GO-ID: 6796]	3,01E-80	Age-dependent response to reactive oxygen species during chronological cell aging [GO-ID: 1320]	6,68E-03
phosphorus metabolic process [GO-ID: 6793]	3,01E-80	oxidation reduction [GO-ID: 55114]	6,90E-03

a. The description of the gene ontology biological processes and the corresponding gene ontology identifiers are given.
b. p-Value calculated as an exponential function.

Table 2. The top 10 of significant biological process of normal and HIV-1 infected macrophages networks [a].

AKT3 via JNK interacts with NFTA and Jun that are targets for the HIV-1 macrophage integration network and are included in the mitogen-activated protein kinase (MAPK) cascade which perform essential functions such as proliferation, survival and inflammation, apoptosis in all cell types. This pathway is associated with others that include the phosphatidylinositol signaling system, Wnt signaling pathway, ERK5 pathway, P53 signaling pathway. (61-63). According with these previous data, we propose that, when AKT3 is turned off by HIV-1 integration, the cross talk with others is disrupted leading to a signaling dysfunction of metabolic associated pathways. When AKT3 was inactive the direct interaction with MKK7 produce a disruption of JNK and after with JUN that would result in a non activation by phosphorilation of apoptotic and cell cycle process. On the other hand inactivation of the MAPK pathway in both macrophages and dendritic cells leads to inhibition of proinflammatory cytokine secretion, downregulation of co-stimulatory molecules such as CD80 and CD86, and ineffective T cell priming. The net result is an impaired innate and adaptive immune response (64, 65).

Recently it have been reported that HIV-1 infection triggers the activation of the PI3K/Akt cell survival pathway in primary human macrophages as reflected by decreased PTEN protein expression and increased Akt kinase activity and renders these cells resistant to cytotoxic insults (54, 61, 64, 65). As result of HIV-1 integration close to AKT3, PTEN, AKT1 and 2, FOXO 1 and MDM2 that are included into the macrophage gene network, would expected a disruption of the apoptotic process.

6. Conclusions

We can conclude that a general effect of HIV-1 integrations in macrophages DNA is to disrupt several signaling pathways that control the normal cell homeostasis. Comparison between normal and infected macrophages of top 10 GO function categories showed the dramatic change of one non-infected macrophage whose main cellular functions are devoted to maintain a cell signaling crucial functions, to one infected in which the most important function are macromolecular biosynthetic process, maintenance of fidelity during DNA-dependent DNA replication, mismatch repair, age-dependent response to reactive oxygen species during chronological cell aging and oxidation reduction. As HIV infected macrophage is an abnormal reservoir in which the metabolic cascades are altered, it is possible to propose that the metabolism of macrophage adapt to perform survival functions where the apoptotic process is interrupted and a SOS metabolism make that the macrophage change of its life style

In silico studies are based upon statistical calculations which permit the drawing of generalizations about a biological process; however since some variables could affecting the *in toto* process, in order to get a real history of Lentivirus integration it would be important to consider that there is another factors, including physiological process and cellular compartments that would be influencing the *in vivo* integration site selection. Some of these are cell-cycle phase, the transcriptional state of the cell, the topology of chromosomal DNA, cell type infected, and presence of co-helper molecules during the PIC complex conformation

[26] Hematti P, Hong BK, Ferguson C, Adler R, Hanawa H, Sellers S, Ingeborg E (2004) Distinct Genomic Integration of MLV and SIV Vectors in Primate Hematopoietic Stem and Progenitor Cells. PLoS Biol. 2: E423

[27] Rick SM, Beitzel BF, Schroder AR, Shinn P, Chen H, Berry CC, Ecker JR, Bushman FD (2004) Retroviral DNA integration: ASLV, HIV, and MLV show distinct target site preferences. PLoS Biol .2: 1127-1137

[28] Maxfield L, Fraize C, Coffin JM (2005) Relationship between retroviral DNA-integration-site selection and host cell transcription. Proc natl. acad. Sci USA. 102: 1436-1441

[29] Soto J, Peña A, Salcedo M, Domínguez MC, Sánchez A, García-Vallejo F (2010) Caracterización Genómica de la Integración In vitro del VIH-1 en células mononucleares de sangre periférica, macrófagos, y células T de Jurkat. Infectio. 14: 20-30

[30] Schroder AR, Shinn P, Chen H, Berry C, Ecker JR, Bushman F (2002) HIV-1 integration in the human genome favors active genes and local hotspots. Cell. 110: 521–529

[31] Derse D, Crise B, Li Y, Princler G, Stewart C, Connor F, Hughes H, Munroe D, Wu X (2007) HTLV-1 integration target sites in the human genome: comparison with other retroviruses. J. virol. 81: 6731-6741

[32] Felice B, Cattoglio C, Cittaro D, Testa A, Miccio A, Ferrari G, Luzi L, Recchia A, Malivio, F (2009) Transcription Factors Binding Sites Are Genetic Determinants of Retroviral Integration in the Human Genome. PLoS One. 4: e4571

[33] MacNeil A, Sankale JL, Meloni S, Sarr A, Mboup S, Kanki P (2006) Genomic Sites of Human Immunodeficiency Virus Type 2 (HIV-2) Integration: Similarities to HIV-1 In Vitro and Possible Differences In Vivo. J. virol. 80: 7316–7321

[34] PARIS CONFERENCE. Supplement (1975) Standardization in human cytogenetic. Cytogenet. Cell Genet. 1971; 15: 203-238

[35] STATSOFT, INC: STATISTICA (data analysis software system), 2004, version 7. Available on: www.statsoft.com

[36] Carteau S, Hoffmann C, Bushman F (1998) Chromosome structure and human immunodeficiency virus type 1 cDNA integration: centromeric alphoid repeats are a disfavored target. J. virol. 72: 4005–4014.

[37] Taganov KD, Cuesta I, Daniel R, Cirillo LA, Katz RA (2004) Integrase specific enhancement and suppression of retroviral DNA integration by compacted chromatin structure in vitro. J. virol. 78: 5848–5855.

[38] Lander ES, Linton LM, Birren B, Nusbaum C, Zody MC, Baldwin J, Devon K, Dewar K, Doyle M, FitzHugh W, Funke R, Gage D, Harris K, Heaford A (2004) International Human Genome Sequencing Consortium. Finishing the euchromatic sequence of the human genome. Nature. 431: 931-945

[39] Zodyl MC, Garber M, Adams D, Sharpe T, Harrow J, Ames R, Nicholson C (2006) DNA sequence of human chromosome 17 and analysis of rearrangement in the human lineage. Nature. 440: 1045-1049

[40] Albanese A, Arosio D, Terreni M, Cereseto A (2008) HIV-1 Pre-Integration Complexes Selectively Target Decondensed Chromatin in the Nuclear Periphery. PLoS one. 3: e2413. doi:10.1371/journal.pone.0002413

[41] Fu W, Sanders-Beer BE, Katz KS, Maglott DR, Pruitt KD, Ptak RG (2009) Human immunodeficiency virus type 1, human protein interaction database at NCBI. Nucleic acid. res. 37(Database issue):D417-22

[42] Song, G, Ouyang G, Bao S (2005) The activation of Akt/PKB signaling pathway and cell survival. J. cell. mol. med. 9 (1): 59-71.

[43] Balakrishnan S, Tastan O, Carbonell J, Klein-Seetharaman J (2009) Alternative paths in HIV-1 targeted human signal transduction pathways. BMC Genomics. 10 (Suppl. 3):S30.

[44] Fu W, Sanders-Beer BE, Katz KS, Maglott DR, Pruitt KD, Ptak RG (2009) Human immunodeficiency virus type 1, human protein interaction database at NCBI. Nucleic acids res. 37: D417–D422.

[45] Perelson AS, Neumann AU, Markowitz M, Leonard JM, Ho DD (1996) HIV-1 dynamics in vivo: virion clearance rate, infected cell lifespan, and viral generation time. Science. 271 (5255):1582–1586.

[46] Sirskyj D, Thèze J, Kumar A, Kryworuchko M (2008) Disruption of the cc cytokine network in T cells during HIV infection. Cytokine. 43 (1): 1–14.

[47] von Schwedler UK, Stuchell M, Müller B, Ward DM, Chung HY, Morita E, Wang HE, Davis T, He GP, Cimbora DM, Scott A, Kräusslich HG, Kaplan J, Morham SG, Sundquist WI (2003) The protein network of HIV budding. Cell. 114 (6): 701–713.

[48] Bandyopadhyay S, Kelley R, Ideker T (2006) Discovering regulated networks during HIV-1 latency and reactivation. Pac. symp. biocomput. 354–366.

[49] Chun T, Carruth LM, Finzi D (1997) Quantification of latent tissue reservoirs and total body viral load in HIV-1 infection. Nature. 387:83–188.

[50] Bagnarelli P, Valenza S, Menzo R, Sampaolesi PE, Varaldo L, Butini M, Montoni CF, Perno S, Aquaro A, Mathez D (1996) Dynamics and modulation of human immunodeficiency virus type 1 transcripts in vitro and in vivo. J. virol. 70 (11):7603–7613.

[51] Perelson AS, Neumann AU, Markowitz M, Leonard JM, Ho DD (1996) HIV-1 dynamics in vivo: virion clearance rate, infected cell lifespan, and viral generation time. Science. 271(5255):1582–1586.

[52] Gendelman HE, Orenstein JM, Baca LM, Weiser B, Burger H, Kalter DC, Meltzer MS (1989) The macrophage in the persistence and pathogenesis of HIV infection. AIDS. 3:475–495.

[53] Herbein G, Gras G, Khan KA, Abbas W (2010) Macrophage signaling in HIV-1 infection. Retrovirology. 9:7–34.

[54] Shannon P, Markiel A, Ozier O, Baliga NS, Wang JT, Ramage D, Amin N, Schwikowski B, Ideker T (2003) Cytoscape: a software environment for integrated models of biomolecular interaction networks. Genome res. 13 (11):2498–2504.

[55] Saeed A, Bhagabati NK, Braisted JC, Liang W, Sharov W, Howe V, Li J, Thiagarajan M, White JA, Quackenbush J, (2006) TM4 microarray software suite. Methods enzymol. 411:134–193.

[56] Maere S, Heymans K, Kuiper M (2005) BiNGO: a cytoscape plug-in to assess overrepresentation of gene ontology categories in biological networks. Bioinformatics. 21 (16):3448–3449.

[57] Ideker T, Ozier O, Schwikowski B, Siegel AF (2002) Discovering regulatory and signalling circuits in molecular interaction networks. Bioinformatics. 18:S233–S240.

[58] Maxfield L, Fraize C, Coffin JM (2005) Relationship between retroviral DNAintegration-site selection and host cell transcription. Proc natl acad.sci. 102 (5):1436–1441.

[59] Schroder AR, Shinn P, Chen H, Berry C, Ecker JR, Bushman F (2002) HIV-1 integration in the human genome favors active genes and local hotspots. Cell. 110 (4):521–529.

[60] Dérijard B, Hibi M, Wu IH, Barrett T, Su B, Deng T, Karin M, Davis RJ (1994) JNK1: a protein kinase stimulated by UV light and Ha-Ras that binds and phosphorylates the c-Jun activation domain. Cell. 76:1025–1037.

[61] Mordret G, (1993) MAP kinase: a node connecting multiple pathways. Biol. cell. 79:193–207.

[62] Rao KM, (2001) MAP kinase activation in macrophages. J. Leukoc. Biol. 69 (1):3–10.

[63] Osaki M, Oshimura M, Ito M (2004) PI3K-Akt pathway: its functions and alterations in human cancer. Apoptosis. 9 (6):667–676.

[64] Chugh P, Bradel-Tretheway B, Monteiro-Filho CM, Planelles V, Maggirwar SB, Dewhurst S, Kim B (2008) Akt inhibitors as an HIV-1 infected macrophagespecific antiviral therapy. Retrovirology. 5:11.

Sequence Analysis

Chapter 4

SeqAnt 2012: Recent Developments in Next-Generation Sequencing Annotation

Matthew Ezewudo, Promita Bose, Kajari Mondal, Viren Patel, Dhanya Ramachandran and Michael E. Zwick

Additional information is available at the end of the chapter

http://dx.doi.org/10.5772/53339

1. Introduction

The discovery of genome-wide genetic variation was central to the field of genomics [1,2]. Now, recent advances in second-generation sequencing technologies and better methods of targeted enrichment mean the detection of genome-wide patterns of genetic variation will soon be a routine operation [3,4]. Yet these advances in DNA sequencing have revealed a new bottleneck: the functional classification and interpretation of newly discovered genetic variation.

The scale of this problem is enormous. The high throughput and low cost of second-generation sequencing platforms now allow geneticists to routinely perform single experiments that identify tens of thousands to millions of variant sites in a single individual, but the methods that exist to annotate these variant sites using information from publicly available databases are too slow to be useful for the large sequencing datasets being generated. Because sequence annotation of variant sites is required before functional characterization can proceed, the lack of a high-throughput pipeline to annotate variant sites efficiently can be a major bottleneck in genetics research and clinical applications of genomics technologies.

To address this problem, we developed the Sequence Annotator (SeqAnt, http://seqant.genetics.emory.edu/), an open source web service and software package that rapidly annotates DNA sequence variants and identifies recessive or compound heterozygous loci in human, mouse, fly, and worm genome sequencing experiments [5]. Variants are characterized with respect to their functional type, frequency, and evolutionary conservation. Annotated variants can be viewed on a web browser, downloaded in a tab-delimited text file, or directly uploaded in a Browser Extensible Document (BED) format to the UCSC Genome Browser. To demonstrate the speed of SeqAnt, we annotated a series of

2.2. SeqAnt 2.0 - Binary database upgrades

One of the unique features of SeqAnt is the ease and speed with which variant information is accessed from a set of customized binary databases. The SeqAnt binary databases are created from flat text table files obtained from the UCSC Genome Browser website [6]. Five main types of data constitute the SeqAnt binary databases. These include:

1. Reference Genome Sequence
2. RefGene Annotation
3. dbSNP Variation Data
4. PhastCons Evolutionary Conservation Scores
5. PhyloP Evolutionary Conservation Score

Standard queries, implemented through the web interfaces described above, are able to extract the annotation information from the binary databases. The actual structure of the binary databases is not directly visible to a SeqAnt user, but is worth examining in greater detail. The Reference Genome Sequence provides the basic backbone for other annotation information. Reference sequences for a given species are organized by different builds (i.e. human genome 18, human genome 19). Within each build, data are organized by chromosome, which reflects the structure of the flat files obtained from UCSC. The RefGene Annotation is the collection of information pertaining to known genes for a given species and build. This information is also organized by chromosome. The collection of variant sites in a given species is contained within the dbSNP Variation Data that is also organized by chromosome. Finally, the SeqAnt 2.0 binary databases include two different measures of evolutionary conservation for all sites in a given reference genome sequence. The PhastCons score is best used to detect functional elements in noncoding sequences, whereas the phyloP score provides a measure of the evolutionary conservation of single sites and is most useful for evaluating sites located in coding regions of genes.

Binary files are significantly smaller than their corresponding flat files, so querying binary files uses less memory than the same analysis performed with a flat file. Considering the vast amount of data that has to be accessed during sequence annotation of large genomic regions, the significant difference in the size of the binary files versus flat files helps to account for the speed with which information is processed using binary files. SeqAnt 2.0 updated a number of these specific binary files; a detailed description of the changes follows in the next sections.

2.2.1. Upgrade of dbSNP to SNP132 Track for hg19 Assembly (Homo sapiens)

The original goal of the dbSNP database (http://www.ncbi.nlm.nih.gov/projects/SNP/) was to develop a comprehensive catalog of common (>5% frequency) human genetic variation [13,14]. These variants were subsequently validated by genotyping in multiple human populations, and their patterns of statistical correlation among variants, known as linkage disequilibrium, were revealed in the HapMap project [15,16]. SeqAnt 1.0 included data from the SNP131 track from the dbSNP [17]. SeqAnt 2.0 was updated to the SNP132 build, which was characterized and uploaded to the UCSC Genome Browser in the summer of 2011. SNP132 has an expanded collection of variant sites that can help researchers determine whether an identical variant has been seen before in a different individual.

Results Directory
--Compound Replacement SNP
--Summary.txt
--Log

All_variations

--Exonic Indel
--Exonic Replacement SNP
--Exonic Silent SNP
--Intergenic Indel
--Intergenic SNP
--Intronic Indel
--Intronic SNP
--UTR Indel
--UTR SNP

Bed_Annotations

--Exonic Indel bed
--Intergenic Indel bed
--Intergenic SNP bed
--Intronic Indel bed
--Intronic SNP bed
--Replacement SNP bed
--Silent SNP bed
--UTR Indel bed
--UTR SNP bed
--UCSC bed

Unique_variations

--Unique Exonic Indel
--Unique Exonic Replacement SNP
--Unique Exonic Silent SNP
--Unique Intergenic Indel
--Unique Intergenic SNP
--Unique Intronic Indel
--Unique Intronic SNP
--Unique UTR Indel
--Unique UTR SNP

Figure 5. Contents of SeqAnt Output Directory. Directories are in bold; individual files shown in a standard font face.

[33] Sun M, Mondal K, Patel V, Horner VL, Long AB, Cutler DJ, Caspary T, Zwick ME. 2012. Multiplex Chromosomal Exome Sequencing Accelerates Identification of ENU-Induced Mutations in the Mouse. *G3 (Bethesda, Md)* 2: 143-150.

High-Performance Computing

Towards a Hybrid Federated Cloud Platform to Efficiently Execute Bioinformatics Workflows

Hugo Saldanha, Edward Ribeiro, Carlos Borges, Aletéia Araújo, Ricardo Gallon, Maristela Holanda, Maria Emília Walter, Roberto Togawa and João Carlos Setubal

Additional information is available at the end of the chapter

http://dx.doi.org/10.5772/50289

1. Introduction

Current generation of high-throughput DNA sequencing machines [1, 35, 66] can generate large amounts of DNA sequence data. For example, the machine HiSeq 2000 from the company Illumina, a current workhorse of genome centers, is capable of generating 600 Giga base-pairs of sequence in one single run [35]. The Human Microbiome project (https://commonfund.nih.gov/hmp) and the 1000 Genomes project (http://www.1000genomes.org) are two examples of projects that are generating terabyte-scale amounts of DNA sequence.

Such vast amounts of data can only be handled by powerful computational infrastructures (also known as cyberinfrastructures), sophisticated algorithms, efficient programs, and well-designed boinformatics workflows. As a response to this challenge, a large ecosystem composed by different technologies and service providers has emerged in recent years with the paradigm of cloud computing [2, 58, 63, 71]. In this paradigm users have transparent access to a wide variety of distributed infrastructures and systems. In this environment, computing and data storage necessities are accomplished in different and unanticipated ways to give the user the illusion that the amount of resources is unrestricted.

In this scenario, cloud computing is an interesting option to control and distribute processing of large volumes of data produced in genome sequencing projects and stored in public databases that are widespread in distinct places. However, considering the constant growing of computational and storage power needed by different bioinformatics applications that are continously beeing developed in different distributed environments, working with one single cloud service provider can be restrictive for bioinformatics applications. Working with more than one cloud can make a workflow more robust in the face of failures and unanticipated needs. Cloud federation [11, 14, 15] is one such solution. Cloud federation offers other advantages over single-cloud solutions. Bioinformatics centers can profit from participation in a cloud federation, by having access to other center programs, data, execution and

storage capabilities, in a collaborative environment. The federation can abstract cloud-specific mechanisms, thus potetially making the use of such a resource more user-friendly and easier to install and customize. This is particularly valuable for small and medium centers that can enlarge their hardware resources and software tools using machines and programs of other centers integrating a federated system.

In this work, we propose a hybrid federated cloud computing platform that aims at integrating and controlling different bioinformatics tools in a distributed, transparent, flexible and fault tolerant manner, also providing highly distributed processing and large storage capability. The objective is to make possible the use of tools and services provided by multiple institutions, public or private, that can be easily aggregated to the cloud. We also discuss a use case of this platform, a bioinformatics workflow for identifying differentially expressed genes in cancer tissues.

2. Federated cloud computing

There are many distinct definitions of cloud computing. According to [29], cloud computing could be defined as *"a computational paradigm highly distributed, directed by a scale economy, in which the computational power, storing, abstract platforms and services, virtualized, managed and dinamically scalable are provided on demand by external users through the Internet".*

[71], using all the characteristics collected from the literature, proposed a definition of clouds as *"a big pool of virtualized resources, easily usable. The resources can be reconfigured dinamically according to a variable load, allowing optimized using. This pool is typically explored by a pay-per-use model in which guarantees are offered by the infrastructure provider, following a service contract".* These authors attempted to define cloud computing using only common characteristics in cloud providers, but they did not find features that were mentioned by all providers. The most common were scalability, pay-per-use model and virtualization.

From these definitions we can state that the goal of cloud computing is to offer to users the idea that they have unrestricted resources, but they have to pay only for those effectively used (model pay-per-use). Another significant advantage of clouds is the management of the computational infrastructure, relieving users from concerns such as power failures and backups. The property of allocating computational resources depending on user needs is called elasticity.

Cloud services can be deployed by providers in different ways [48]:

- Private cloud: operated for the use of a single organization. It can be managed by the organization itself or by external ones.

- Community Cloud: shared by several organizations and used as a tool for a specific group of users with common interests.

- Public Cloud: available to the general public or a large corporate group that is part of the organization that sells this service.

- Hybrid cloud: composed of two or more clouds (private, community or public) that remain separate entities, but that are bound together by standardized or proprietary technologies that enable portability of data and applications.

In clouds, one of the key technologies adopted to execute bioinformatics programs is the Apache Hadoop framework [6], in which the MapReduce [25] model and its distributed file system (HDFS) [13] are used as infrastructure to distribute large scale processing and data storage. In the MapReduce model parallelization does not require communication among simultaneously processed tasks, since they are independent from one another.

Bittman [11] claimed that the evolution of cloud computing market could be divided in three phases. In phase 1 (Monolithic), cloud computing services were based on proprietary architectures, or cloud services were delivered by megaproviders. In phase 2 (Vertical Supply Chain), some cloud providers leveraged services from other providers, i.e. independent software vendors (ISVs) developed applications as a service using an existing cloud infrastructure. Clouds were still proprietary, but ecosystems construction started. In phase 3 (Horizontal Federation), smaller providers would horizontally federate to gain economy of scale and efficient use of their assets. Projects would leverage horizontal federation to enlarge their capacibilities, more choices at each cloud computing layer would be provided, and discussion about standards would begin.

In general, cloud computing intends to increase efficiency in service delivery, dealing with services including infrastructure, platforms and software, and treating with distinct users like a single user, other clouds, academic institutions and large companies. Besides public clouds maintained by large organizations, hundreds of smaller heterogeneous and independent clouds, private or hybrid, are being developed. In this scenario, cloud federation becomes an interesting way to optimize the use of the resources offered by various organizations. In particular, in this chapter, we are interested in horizontal cloud federation, also called federated cloud computing, inter-cloud [14] or cross-cloud [15].

Federated cloud computing can be defined as a set of cloud computing providers, public and private, connected through the Internet [14, 15]. Among its objectives we distinguish the seemingly availability of unrestricted resources, independence of a single infrastructure provider, and optimization when using a set of distinct resource providers.

Thus, federation allows each cloud computing provider to increase its processing and storage capabilities by requesting more resources to other clouds in the federation when needed. This means that a local cloud provider is able to satisfy user requests beyond its capabilities, since idle resources from other providers can be used. Furthermore, if a provider fails, resources can be requested to another one, providing more fault tolerance.

Although the advantages of federated cloud computing are obvious, its implementation is not trivial, since the participating clouds present heterogeneous and frequently changing resources. Therefore, traditional models of federation are not useful [15]. Typically, federated models are based on *a priori* agreements among their members, noting that these agreements can be inappropriate according to the particular characteristics of a cloud provider. Thus, to make possible the creation of a federated cloud environment, it is necessary to achieve the following requirements [14, 15]:

- **Automatism**: a cloud member of the federation, using discovery mechanisms, should be able to identify the other clouds in the federation together with their resources, responding to changes in a transparent and automatic way;

- **Application behavior prediction**: the system implementing the federation has to be able to predict demands and behaviors of the offered applications, so that its load balancing mechanism can have its efficiency improved;

- **Mapping services to resources**: the services offered by the federation must be mapped to available resources in a flexible manner so that it can achieve the highest levels of efficiency and cost/benefit. In other words, the schedule must choose the best hardware-software combination to ensure the quality of service at lowest cost, taking into account the uncertainty of the availability of resources;

- **Interoperable security model**: federation must allow the integration of different security technologies so that a cloud member does not need to change its security policies when entering the federation;

- **Scalability in monitoring components**: considering the possible large number of participants, the federation must be able to handle multiple task queues and the largest number of requests, so that management can guarantee that the various cloud providers of the federation will mantain scalability and performance.

It is noteworthy that issues to choose an appropriate cloud provider and lack of common cloud standards hinder the interoperability across these federated cloud providers. Thus, nowadays the user is faced with the challenging problem of selecting the appropriate cloud that fits his or her needs. To address this problem, the BioNimbus platform offers to users a federated platform that can execute bioinformatics applications in a transparent and flexible manner. This is possible because BioNimbus offers standardized interfaces and intermediate services to manage the integration of different cloud providers. Moreover, as will be seen next, BioNimbus was designed to incorporate the requirements defined by [15].

3. BioNimbus: a federated cloud platform

As mentioned before, cloud computing is a promising paradigm for bioinformatics due to its ability to provide a flexible computing infrastructure on demand, its seemingly unrestricted resources, and the possibility to distribute execution in a large number of machines leading to a significant reduction in processing time due to the high degree of achieved parallelism. Some bioinformatics tools have been implemented in cloud environments belonging to several infrastructures of physically separated institutions, which makes it difficult for them to be integrated.

BioNimbus [12, 62] is a federated cloud platform designed to integrate and control different bioinformatics tools in a distributed, flexible and fault tolerant manner, providing rapid processing and large storage capabilities, transparently to users. BioNimbus joins physically separate platforms, each modelled as a cloud, which means that independent, heterogenous, private/public clouds providing bionformatics applications can be used as if they were a single system. In BioNimbus, resources of each cloud can be maximally explored, but if more are required, other clouds can be requested to participate, in a transparent manner. BioNimbusis thus able to satisfy further service allocation requests sent by its users. The objective is to offer an environment with apparently unrestricted computational resources given that computing and storage space demands are always provided on demand to the users.

3.1. BioNimbus architecture

All the components of BioNimbus architecture together with their funcionalities are defined such that it allows simplicity, speed and eficiency when a new cloud provider enters in the federation. Another key characteristic is the communication among the BioNimbus components that is realized through a Peer-to-Peer (P2P) [67] network, guaranteeing the following properties:

- Fault tolerance, since there is not a single fail point. Thus, even if some nodes fail, the others can work;

- Efficiency, since there is not a single bottleneck. Then, messages are end-to-end and not routed by a single node;

- Flexibility, since clouds can operate independently or in a coordinated manner;

- Scalability, since the use of a P2P network allows integration of thousands of interconnected machines.

BioNimbus (Figure 1) architecture enables the integration of different cloud computing platforms, meaning that independent, heterogeneous, private or public providers may offer their bioinformatics services in an integrated manner, while maintaining their particular characteristics and internal policies. BioNimbus is composed of three layers: application layer, core layer and cloud provider layer.

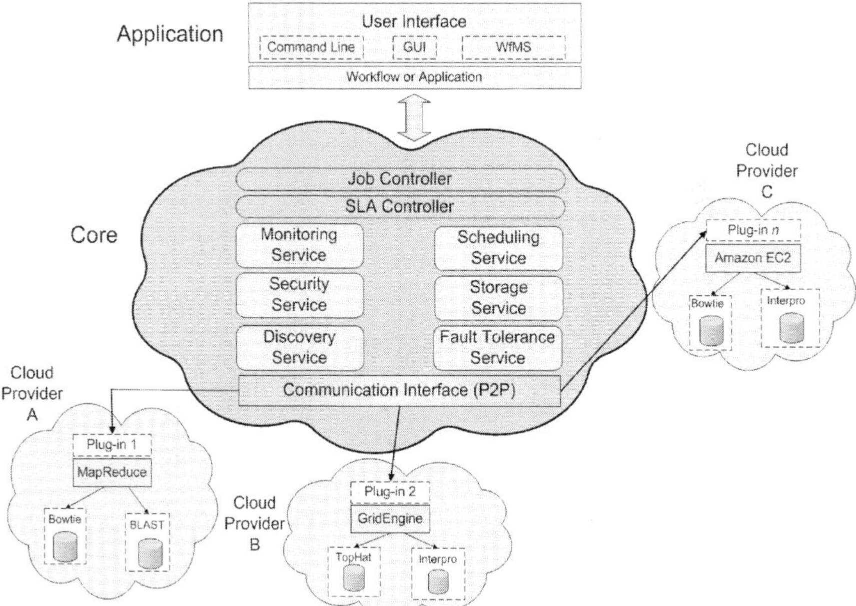

Figure 1. The architecture of BioNimbus hybrid federated cloud.

3.1.1. Application layer

This layer provides the service of interaction with users, which can be implemented by web pages, command lines, graphical interfaces (GUI) or workflow management systems (WfMS). Users can execute workflows or a single application, choosing among available services. Job controller service has the function of collecting user requests and sending the input data to the core layer. Moreover, this layer is responsible for showing to each user the current status of his running applications. Users can list the files stored in the federation, upload or download files, create and execute workflows in the BioNimbus cloud environment.

3.1.2. Cloud provider layer

This layer encompasses the cloud providers belonging to BioNimbus. The previous described core layer ensures a unified view of the cloud, which allows users to see all the resources available on each cloud as if they were one unified system.

A plug-in service is used to integrate a cloud provider (public or private) in the federation. Each plug-in service is an interface that aims at communicating the BioNimbus core with each cloud provider. Cloud providers can communicate among themselves also using the core layer. Each plug-in needs to map the requests sent by the core components to the corresponding actions that have to be realized in each cloud provider. This implies that each cloud requires a special plug-in service. Furthermore, to integrate distinct providers (public or private), each plug-in needs to treat three different kinds of requests: information about the provider infrastructure, task management and file transfer.

3.2. Core layer

This layer is responsible for managing the federated environment. Among their functions are: identification of new providers with their corresponding hardware and software resources; task scheduling and controlling; definition, establishment and monitoring of SLA (Service Level Agreement); storage and managing of input and output files; maintainance of the online environment; and the election of new coordinators for each requested service. To each function implemented in this layer, a controller service was included in the architecture, as described next.

3.2.1. Discovery service

This service identifies the cloud providers integrating the federation, and consolidates information about storage and processing capabilities, network latency, availability of resources, available bioinformatics tools, details of parameters and input and output files. To realize this, the discovery service waits for information published by providers about their infrastructure and available tools. To consolidate these data, the discovery service maintains a data structure that is updated whenever new data is received. Furthermore, the discovery service has a policy of controlling each provider, removing from the federation those providers not regularly sending updated information, which guarantees the correct and update task execution on the federated cloud. Regarding to the entrance of a cloud provider in BioNimbus, *a priori*, at any time a peer participating in the P2P network can start the process of publishing its resources (storage and processing capabilities) and available bioinformatics applications. However, for security and controlling purposes, permission to join the federation as a provider must be verified with the support of the security service whenever any information about a new provider arrives to the discovery service.

As can be seen from the above description, an efficient resource discovery mechanism plays a central role in our federated cloud, since the information gathered by this service is essential to other services to properly perform their functions. According to [49], in large-scale distributed services, a resource discovery infrastructure has to meet the following key requirements: it must be scalable so it can handle thousands of machines without being unavailable or losing performance; it must be able to handle both static and dynamic resources; and it must be flexible enough so its queries could be extended in order to handle different types of resources. Possible implementations of a resource discovery service could be developed using central or hierarchical approaches, but these are known to have serious limitations of scalability, fault-tolerance and network congestion [56].

In BioNimbus, we plan to use a publish/subscribe mechanism, in which providers publish information about their resources to a decentralised resource discovery system. This system will use a Distributed Hash Table (DHT) data structure [7] in order to achieve low management costs and network overhead, efficient resource searching and fault-tolerance. For resource information handling, we plan to use serializable and extendable formats such as the JSON format [23]. In this way it will be possible for the federated cloud to deal with different types of information, thus causing the least impact possible.

3.2.2. Job controller

The job controller links the core and the application layers of BioNimbus. It first calls the security service to verify if a user has permission (authentication) to execute jobs in BioNimbus and what are the credentials of this user. Moreover, the job controller's main function is to manage distinct and simultaneously running workflows, noting that the workflows may belong to the same or to different users. Thus, for each accepted workflow, the job controller generates an associated ID and controls each workflow execution using this ID.

3.2.3. SLA controller

According to [74], SLA is a formal contract between service providers and consumers to guarantee that consumers' service quality expectations can be achieved. In BioNimbus, the SLA controller is responsible for implementing the SLA lifecycle, which has six steps: discovers service providers, defines SLA, establishes agreement, monitors SLA violation, terminates SLA and enforces penalties for violation. A SLA template represents, among others, the QoS parameters that a user has negotiated with BioNimbus. The user can populate, through an user interface, a suitable template with required values or even define a new SLA from scratch in order to describe functional (e.g. CPU cores, memory size, CPU speed, OS type and storage size) and non-functional (e.g. response time, budget, data transfer time, availability and completion time) service requirements. In bioinformatics, functional requirements are number of cores, amounts of memory and storage, CPU speed, bioinformatics programs and databases and respective versions. Non-functional requirements are latency (transfer rate) and uptime (reliability, in the sense that it measures how frequently a cloud provider is running tasks or if it is not entering and leaving the federation).

The SLA controller has the responsibility to investigate whether the SLA template submitted by the user can be supported by the federated cloud platform. For this, the SLA controller creates a strong interdependency among the clouds. We intend to use a Single Sign-On (SSO) protocol [52] so that no central authority is in charge of its users' authentication, which prevents a single point of failure and allows scalability according to the number of users. We chose the OpenID standard [57] as our SSO mechanism, since it has been used by corporate and academic sites around the world. OpenID allows each "site" (e.g. a cloud) to provide an authentication facility to its users so that they do not need to authenticate with each other cloud integrating the federation. Instead, each cloud provider acts as an identity provider for user credentials, so that each user should authenticate with its affiliated provider. Once this user is authenticated, each time his/her credentials are required, OAuth [10] allows a user's site to forward authorization without exposing the user account or login information.

- **Authorization**: The authorization of a federated cloud resource is provided by the Access Control Lists (ACLs) [69] provided by each cloud provider. An ACL determines who can access a given resource, e.g. disk storage, CPU cycles and bioinformatics services. Therefore, each cloud is able to determine access patterns so that it can control its resource's uses.

- **Confidentiality**: Communication between each two cloud providers is established using TLS/SSL [68] connection. The use of secure connections between two clouds in the federation is not enforced by our model, but it can be provided as well. As far as we know, few cloud systems provide secure intra-cloud communication. Each cloud should provide a certificate that will be used by hosts in two clouds to establish a secure connection. As we improve BioNimbus, we plan to include audit trails so that each required resource can be available when needed.

3.2.7. Fault tolerance service and high availability

This service guarantees that all the core services are always available. In a cloud environment, machine failures occur, and it is well known among the cloud community that those failures are the norm rather than the exception. Thus, any federated cloud should be designed for fault recovering and system availability. Therefore, a fault tolerance service is an essential part of our federated cloud, and has the objective of providing high availability and resiliency against periodic or transient failures.

There are extensive studies in the literature on failure detection systems [16, 31, 45, 70]. On the other hand, few systems are designed to scale with a large number of nodes as those found on clouds. Thus, an important requirement of our fault detection service is to be scalable with a large number of machines. We adopted a modified gossip based failure detector proposed by Renesse et al [70], which works as described. Each host runs the gossip failure detector service, which maintains a list of known hosts in the cloud. Every T_{seconds}, a host increases a heartbeat, and at random chooses a set of nodes for sending a list of known nodes. When received, each list is merged with the host current list, assuming the largest heartbeat for each node in the list. If a node does not update its heartbeat for a T_{elapsed} time, then it will be marked as failed. Note that a node may be marked as failed due to slow network links or even in presence of a fractioned network. But our failure detection service is conservative so that it only purges a host from the list after a $T >= 2 * T_{\text{elapsed}}$.

Besides using this gossip based failure detector, we use a coordination service based on atomic broadcast protocol [59]. The open source system Apache Zookeeper [34] runs on each cloud and allows our system to detect node failures and realize an election of leaders among the cloud machines, in order to guarantee the services availability, including discovery and fault tolerance services. Zookeeper is used to elect some of the nodes that are known as gossip servers. Those servers are dinamically chosen so that they can exchange the list of nodes among the cloud providers. This helps to reduce the bandwidth between two clouds to a few servers.

3.2.8. Scheduling service

This service dynamically distributes tasks among the cloud providers belonging to the federation, maintaining a register for the allocated tasks, controlling load of each cloud provider, and redistributing the tasks when resources are overloaded. The scheduling service is responsible for receiving the tasks created from the user requests, and maintaining a record about the status of each executed task. Before being executed in a cloud provider, a task is sent to the scheduling service, which uses one or more scheduling policies to choose the cloud provider that will execute this task, according to the negotiated SLA. Each policy receives a list of tasks to be scheduled and an agreement ID, and returns a mapping of the tasks and the cloud providers where these tasks will be executed. To do this, the scheduling policy communicates with the discovery service. The scheduling policy should consider the SLA QoS parameters and the margin values accepted by the cloud providers (e.g. gold, silver and bronze SLA level). These parameters are important for a matching of a cloud provider that is done by the scheduling policy, and therefore the user needs to give reasonable values for them. Some typical SLA parameters used in context of a cloud provider are CPU cores, CPU speed, memory size, in/ou bandwidth, OS type, storage size, response time, budget, data transfer time, completion time and availability. In BioNimbus, the scheduling service can be easily modified to use different scheduling policies.

We implemented a new *DynamicAHP* algorithm in BioNimbus [12]. The key idea of DynamicAHP is to map available resources of the cloud providers to the requested tasks, then associating a cloud to execute each task. This algorithm is based on a decision making strategy proposed by [61]. DynamicAHP worked well on a first BioNimbus prototype, since it was capable to dinamically scale using only the knowledge about the length of each task input file, while performing load balancing among the cloud providers. Since BioNimbus stores information about the cloud providers such as network latency and wait time in the execution queues, DynamicAHP reduced costs and execution time of the tasks. The promising results obtained from developing DynamicAHP in BioNimbus showed that good scheduling algorithms can really lower the time to execute bioinformatics applications in federated clouds.

It is interesting to investigate new scheduling metrics, mainly related to costs. For example, a public cloud can be associated to a lower priority due to its associated costs, when compared to other public clouds integrating the federation. Another idea is to assign weights to the metrics that could set a priority order among them. To develop and analyze a model capable of storing information about the executed tasks, such that the scheduling service could combine this information to estimate the execution time of a particular task is another challenging project.

Total time (seconds)	File transfer time (seconds)	Percentage of file transfer time related to the total time
4230.297	2114.996	50.0%
4123.492	2264.552	54.9%
4098.571	2337.454	57.0%
4030.492	2297.580	57.0%
3807.501	2229.992	58.6%
3145.645	2168.201	68.9%
3113.729	2116.199	68.0%
3066.488	2058.771	67.1%
3032.701	2018.942	66.6%
3001.165	2137.157	71.2%
2952.875	2087.761	70.7%
2849.506	2074.117	72.8%
2801.489	2023.309	72.2%
2680.382	1892.002	70.6%
2587.076	2006.842	77.6%
2579.184	1959.727	76.0%
2533.254	1928.888	76.1%
2405.470	1899.626	79.0%

Table 2. Total and file transfer times of the longest jobs executed in BioNimbus.

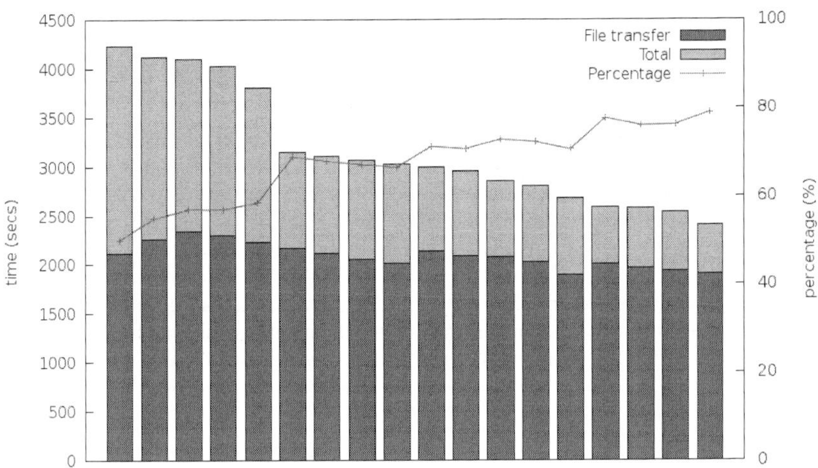

Figure 5. Comparing the total and file transfer times of the longest jobs executed in BioNimbus. The file transfer time is colored red, while its percentage related to the total time is shown in blue.

the input files are all simultaneously downloaded, i.e. there are no priorities for downloads; (iii) jobs are now canceled based only on the wait time in the pending jobs list, i.e. the file transfering time is not considered; and (iv) jobs with small input files that were sent to a cloud provider after jobs with large input files got executed earlier, while the later were still downloading their input data.

Table 3 and Figure 6 show the number of jobs executed in a single cloud provider and on both. Note that, including the transfer time, jobs with smaller inputs execute faster on two cloud providers, since the possibility to cancel delayed jobsthat are running and scheduling them again lowered the total execution time. Besides, when files are small, the time to transfer files is rapid, while when they are large the transfer time strongly affects the total execution time (as shown in Table 2). Thus, for large files, the storage policy has to be very carefully designed using replication and fragmentation in order to significantly decrease file transfer time.

Cloud Providers	until 200 seconds	between 200 seconds and 1000 seconds	above 1000 seconds
University of Brasilia	34	30	32
EC2 Amazon	37	27	32
UnB and EC2	64	8	24

Table 3. Number of executed jobs, where time includes the file transfer time.

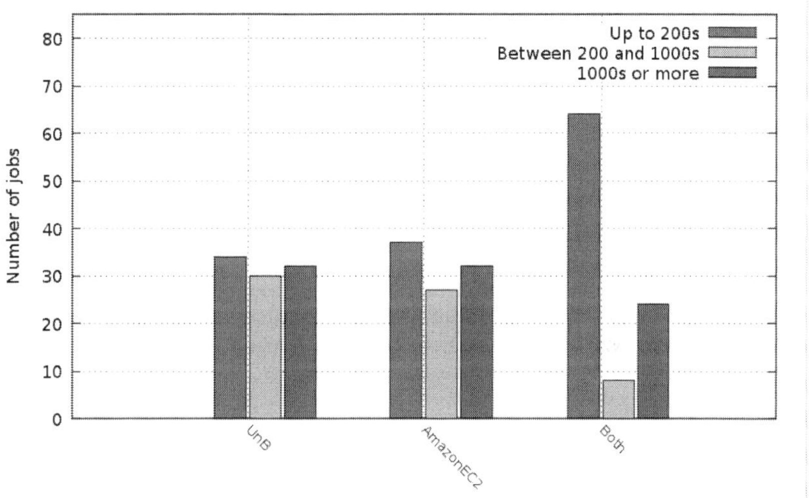

Figure 6. Comparing the number of executed jobs in BioNimbus, where time (in seconds) includes the file transfer time.

5. Related work

In this section, we discuss cloud projects designed to accelerate execution and increase the amount of storage available to bioinformatics applications. When compared to BioNimbus, these projects are dedicated to particular applications or are executed in a single cloud environment. BioNimbus intends to integrate public and private centers offering bioinformatics applications in one single platform using the hybrid federation cloud paradigm.

Author details

Hugo Saldanha, Edward Ribeiro, Carlos Borges, Aletéia Araújo, Ricardo Gallon, Maristela
Holanda and Maria Emília Walter
University of Brasília, Brasil

Roberto Togawa
Embrapa/Genetic Resources and Biotechnology, Brasil

João Carlos Setubal
University of São Paulo, Brasil

7. References

[1] 454 [2012]. 454 life sciences. roche diagnostics corporation, `http://www.454.com/`.

[2] Amazon [2012]. Amazon Elastic Compute Cloud (Amazon EC2), `http://aws.amazon.com/ec2/`.

[3] Angiuoli, S. V., Matalka, M., Gussman, A., Galens, K., Vangala, M., Riley, D. R., Arze, C. & White, J. R. [2011]. Clovr: A virtual machine for automated and portable sequence analysis from the desktop using cloud computing, *BMC Bioinformatics* 12(356): 1–15.

[4] Angiuoli, S. V., White, J. R., Matalka, M., White, O. & Fricke, W. F. [2011]. Resources and costs for microbial sequence analysis evaluated using virtual machines and cloud computing, *PLoS ONE* 6(10): e26624.

[5] Apache [2011]. The Apache Software Foundation: Apache, `http://couchdb.apache.org/`.

[6] Apache [2012]. Apache Hadoop, `http://hadoop.apache.org/`.

[7] Balakrishnan, H., Kaashoek, M. F., Karger, D., Morris, R. & Stoica, I. [2003]. Looking up data in p2p systems, *Commun. ACM* 46(2): 43–48.
URL: *http://doi.acm.org/10.1145/606272.606299*

[8] Bermbach, D., Klems, M., Tai, S. & Menzel, M. [2011]. Metastorage: A federated cloud storage system to manage consistency-latency tradeoffs, *IEEE International Conference on Cloud Computing (CLOUD)*, IEEE, pp. 452–459.

[9] BGI [2012]. Bio-Cloud Computing, `http:www.genomics.cn/en/navigation/show_navigation?nid=4143`.

[10] Bhandari, A. & Singh, M. [2010]. 2010 the oauth 1.0 protocol, internet engineering task force (ietf), rfc:5849.
URL: *http://tools.ietf.org/html/rfc5849*

[11] Bittman, T. J. [2008]. Th evolution of the cloud computing market, `http://blogs.gartner.com/thomas_bittman/2008/11/03/the-evolution-of-the-cloud-computing-market`.

[12] Borges, C. A. L., Saldanha, H. V., Ribeiro, E., Holanda, M. T., Araujo, A. P. F. & Walter, M. E. M. T. [2012]. Task scheduling in a federated cloud infrastructure for bioinformatics applications, *in* INSTICC (ed.), *2nd International Conference on Cloud Computing and Services Science ()*, CLOSER 2012, Porto, Portugal, pp. 1–7.

[13] Borthakur, D. [2008]. The Apache Software Foundation: HDFS Architecture, `http://hadoop.apache.org/common/docs/r0.20.2/hdfs_design.pdf`.

[14] Buyya, R., Ranjan, R. & Calheiros, R. N. [2010]. Intercloud: Utility-oriented federation of cloud computing environments for scaling of application services, *Proceedings of the*

10th International Conference on Algorithms and Architectures for Parallel Processing (ICA3PP 2010), Springer, pp. 21–23.

[15] Celesti, A., Tusa, F., Villari, M. & Puliafito, A. [2008]. How to enhance cloud architectures to enable cross-federation, *3rd IEEE International Conference on Cloud Computing (IEEE Cloud 2010)*, Miami, Florida, USA, pp. 337–345.

[16] Chandra, T. D. & Toueg, S. [1996]. Unreliable failure detectors for reliable distributed systems, *Journal of the ACM* 43: 225–267.

[17] Chang, F., Dean, J., Ghemawat, S., Hsieh, W., Wallach, D., Burrows, M., Chandra, T., Fikes, A. & Gruber, R. [2008]. Bigtable: A distributed storage system for structured data, *ACM Transactions on Computer Systems (TOCS)* 26(2): 1–26.

[18] Chen, W. [2000]. On the quality of service of failure detectors, *IEEE Transactions on Computers* 51: 561–580.

[19] Chodorow, K. & Dirolf, M. [2010]. *MongoDB: the definitive guide*, O'Reilly Media, Inc.

[20] Clayman, S., Galis, A., Chapman, C., Toffetti, G., Rodero-Merino, L., Vaquero, L. M., Nagin, K. & Rochwerger, B. [2010]. Monitoring Service Clouds in the Future Internet, *Towards the Future Internet - Emerging Trends from European Research*, IOS Press.

[21] Codehaus [2012]. Jackson Java JSON-Processor, `http://jackson.codehaus.org`.

[22] Cooper, B., Ramakrishnan, R., Srivastava, U., Silberstein, A., Bohannon, P., Jacobsen, H., Puz, N., Weaver, D. & Yerneni, R. [2008]. PNUTS: Yahoo!'s hosted data serving platform, *Proceedings of the VLDB Endowment*, VLDB Endowment, pp. 1277–1288.

[23] Crockford, D. [2006]. The application/json Media Type for JavaScript Object Notation (JSON), RFC 4627 (Informational).
URL: *http://www.ietf.org/rfc/rfc4627.txt*

[24] da Silva, C. A. R. F. O. [2011]. *Data modeling with NoSQL: How, When and Why*, Master's thesis, University of Porto.

[25] Dean, J. & Ghemawat, S. [2004]. MapReduce: simplified data processing on large clusters, *6th Conference on Symposium on Operating Systems Design & Implementation*, USENIX Association, Berkeley, CA, EUA, pp. 10–10.

[26] DeCandia, G., Hastorun, D., Jampani, M., Kakulapati, G., Lakshman, A., Pilchin, A., Sivasubramanian, S., Vosshall, P. & Vogels, W. [2007]. Dynamo: amazon's highly available key-value store, *SIGOPS Oper. Syst. Rev.* 41(6): 205–220.
URL: *http://doi.acm.org/10.1145/1323293.1294281*

[27] Ekanayake, J., Gunarathene, T. & Qiu, J. [2011]. Cloud technologies for bioinformatics applications, *IEEE Transactions on Parallel and Distributed Systems* 22(6): 998–1011.

[28] Elmroth, E. & Larsson, L. [2009]. Interfaces for placement, migration, and monitoring of virtual machines in federated clouds, *Proceedings of the 2009 Eighth International Conference on Grid and Cooperative Computing*, GCC '09, IEEE Computer Society, Washington, DC, USA, pp. 253–260.
URL: *http://dx.doi.org/10.1109/GCC.2009.36*

[29] Foster, I., Zhao, Y., Raicu, I. & Lu, S. [2008]. Cloud Computing and Grid Computing 360-Degree Compared, *Grid Computing Environments Workshop GCE '08*, pp. 1–10.

[30] Hayashibara, N., Défago, X., & Rami Yared, T. K. [2004]. The phi Accrual Failure Detector, *23rd IEEE International Symposium on Reliable Distributed Systems*, pp. 66–78.

[31] Hayashibara, N., Défago, X., Yared, R. & Katayama, T. [2004]. The ϕ accrual failure detector, *RR IS-RR-2004-010, Japan Advanced Institute of Science and Technology*, pp. 1–16.

[32] Hoffa, C., Mehta, G., Freeman, T., Deelman, E., Keahey, K. a. B. B. & Good, J. [2008]. On the use of cloud computing for scientific workflows, *3rd International Workshop on Scientific Workflows and Business Workflows Standards in e-Science*, SWBES 2008, IEEE Digital Library, pp. 640–645.

[33] Hong, D., Rhie, A., Park, S.-S., Lee, J., Ju, Y. S., Kim, S., Yu, S.-B., Bleazard, T., Park, H.-S., Rhee, H., Chong, H., Yang, K.-S., Lee, Y.-S., Kim, I.-H., Lee, J. S., Kim, J.-I. & Seo, J.-S. [2012]. FX: an RNA-Seq analysis tool on the cloud, *Bioinformatics* 28(5): 721–723. URL: *http://dx.doi.org/10.1093/bioinformatics/bts023*

[34] Hunt, P., Konar, M., Junqueira, F. P. & Reed, B. [2010]. Zookeeper: wait-free coordination for internet-scale systems, *USENIX conference on USENIX annual technical conference*, USENIXATC'10, USENIX Association, pp. 11–11.

[35] Illumina [2012]. Illumina life sciences, `http://www.illumina.com`.

[36] Inc., R. H. [2012]. Netty - the Java NIO Client Server Socket Framework, `http://www.jboss.org/netty`.

[37] Jourdren, L., Bernard, M., Dillies, M.-A. A. & Le Crom, S. [2012]. Eoulsan: A Cloud Computing-Based Framework Facilitating High Throughput Sequencing Analyses., *Bioinformatics (Oxford, England)* . URL: *http://dx.doi.org/10.1093/bioinformatics/bts165*

[38] Khetrapal, A. & Ganesh, V. [2006]. Hbase and hypertable for large scale distributed storage systems, Dept. of Computer Science, Purdue University, http://www.uavindia.com/ankur/downloads/HypertableHBaseEval2.pdf.

[39] Kossmann, D., Kraska, T., Loesing, S., Merkli, S., Mittal, R. & Pfaffhauser, F. [2010]. Cloudy: a modular cloud storage system, *Proceedings of the VLDB Endowment*, VLDB Endowment, pp. 1533–1536.

[40] Krampis, K., Booth, T., Chapman, B., Tiwari, B., Bicak, M., Field, D. & Nelson, K. [2012]. Cloud BioLinux: pre-configured and on-demand bioinformatics computing for the genomics community, *BMC Bioinformatics* 13(1): 42+. URL: *http://dx.doi.org/10.1186/1471-2105-13-42*

[41] Lakshman, A. & Malik, P. [2010]. Cassandra: a decentralized structured storage system, *SIGOPS Oper. Syst. Rev.* 44(2): 35–40. URL: *http://doi.acm.org/10.1145/1773912.1773922*

[42] Langmead, B., Hansen, K. & Leek, J. [2010]. Cloud-scale RNA-sequencing differential expression analysis with Myrna, *Genome Biology* 11(8): R83.

[43] Langmead, B., Schatz, M. C., Lin, J., Pop, M. & Salzberg, S. [2009]. Searching for SNPs with cloud computing, *Genome Biology* 10(11): R134.

[44] Langmead, B., Trapnell, C., Pop, M. & Salzberg, S. [2009]. Ultrafast and memory-efficient alignment of short DNA sequences to the human genome, *Genome Biology* 10(3): R25.

[45] Larrea, M., Fernandez, A., Arevalo, S., Carlos, J. & Carlos, J. [2002]. Eventually consistent failure detectors, *Brief Announcement, 14th International Symposium on Distributed Computing (DISC'2000)*, pp. 326–327.

[46] Li, R., Li, Y., Fang, X., Yang, H., Wang, J., Kristiansen, K. & Wang, J. [2009]. SNP detection for massively parallel whole-genome resequencing, *Genome Research* 19(6): 1124–1132. URL: *http://dx.doi.org/10.1101/gr.088013.108*

[47] Marioni, J. C., Mason, C. E., Mane, S. M., Stephens, M. & Gilad, Y. [2008]. Rna-seq: An assessment of technical reproducibility and comparison with gene expression arrays,

Genome Research 18: 1509–1517.
URL: *http://genome.cshlp.org/cgi/content/full/18/9/1509*

[48] Mell, P. & Grance, T. [2009]. The NIST Definition of Cloud Computing, *National Institute of Standards and Technology* 53(6): 50.
URL: *http://csrc.nist.gov/groups/SNS/cloud-computing/cloud-def-v15.doc*

[49] Oppenheimer, D., Albrecht, J., Patterson, D. & Vahdat, A. [2004]. Scalable wide-area resource discovery, *Technical Report UCB/CSD-04-1334*, EECS Department, University of California, Berkeley.
URL: *http://www.eecs.berkeley.edu/Pubs/TechRpts/2004/5465.html*

[50] Orend, K. [2010]. *Analysis and classification of nosql databases and evaluation of their ability to replace an object-relational persistence layer*, Master's thesis, Fakultät für Infoormatik, Technische Universität München.

[51] Padhy, R., Patra, M. & Satapathy, S. [2011]. Rdbms to nosql: Reviewing some next-generation non-relational databases, *International Journal of Advanced Engineering Science and Technologies* 11(1): 15–30.

[52] Pashalidis, A. & Mitchell, C. J. [2003]. A taxonomy of single sign-on systems, *Information Security and Privacy, 8th Australasian Conference, ACISP 2003*, Springer-Verlag, pp. 249–264.

[53] Pratt, B., Howbert, J. J., Tasman, N. & Nilsson, E. J. [2011]. MR-Tandem: Parallel X!Tandem using Hadoop MapReduce on Amazon Web Services, *Bioinformatics* 8: 1–12.

[54] Quinlan, A. R. & Hall, I. M. [2010]. BEDTools: a flexible suite of utilities for comparing genomic features, *Bioinformatics* 26(6): 841–842.
URL: *http://bioinformatics.oxfordjournals.org/content/26/6/841.abstract*

[55] R Development Core Team [2011]. *R: A Language and Environment for Statistical Computing*, R Foundation for Statistical Computing, Vienna, Austria. ISBN 3-900051-07-0.
URL: *http://www.R-project.org*

[56] Ranjan, R., Chan, L., Harwood, A., Karunasekera, S. & Buyya, R. [2007]. Decentralised resource discovery service for large scale federated grids, *Proceedings of the Third IEEE International Conference on e-Science and Grid Computing*, E-SCIENCE '07, IEEE Computer Society, Washington, DC, USA, pp. 379–387.
URL: *http://dx.doi.org/10.1109/E-SCIENCE.2007.27*

[57] Recordon, D. & Reed, D. [2006]. Openid 2.0: a platform for user-centric identity management, *Proceedings of the second ACM workshop on Digital identity management*, DIM '06, ACM, New York, NY, USA, pp. 11–16.

[58] Redkar, T. [2011]. *Windows Azure Platform*, Vol. 1, 2th edn, Apress, Berkeley, CA, USA.

[59] Reed, B. & Junqueira, F. P. [2008]. A simple totally ordered broadcast protocol, *2nd Workshop on Large-Scale Distributed Systems and Middleware*, LADIS'08, ACM, New York, NY, USA, pp. 2:1–2:6.

[60] Robinson, M. & Oshlack, A. [2010]. A scaling normalization method for differential expression analysis of RNA-seq data, *Genome Biology* 11(3): R25+.
URL: *http://dx.doi.org/10.1186/gb-2010-11-3-r25*

[61] Saaty, T. L. [1990]. How to make a decision: The analytic hierarchy process, *European Journal of Operational Research* 48(1): 9 – 26.

[62] Saldanha, H. V., Ribeiro, E., Holanda, M., Araujo, A., Rodrigues, G., Walter, M. E. M. T., Setubal, J. C. & Davila, A. [2011]. A cloud architecture for bioinformatics workflows,

in L. INSTICC (ed.), *1st International Conference on Cloud Computing and Services Science*, CLOSER 2011, pp. 1–8.

[63] Sanderson, D. [2009]. *Programming Google App Engine: Build and Run Scalable Web Apps on Google's Infrastructure*, 1st edn, O'Reilly Media, Inc.

[64] Schatz, M. C. [2009]. CloudBurst: Highly Sensitive Read Mapping with MapReduce, *Bioinformatics* 25: 1363–1369.

[65] Smith, A., Xuan, Z. & Zhang, M. [2008]. Using quality scores and longer reads improves accuracy of Solexa read mapping, *BMC Bioinformatics* 9(1): 128.

[66] Solid [2012]. Life technologies & applied biosystems, `http://www.appliedbiosystems.com/`.

[67] Stoica, I., Morris, R., Liben-Nowell, D., Karger, D. R., Kaashoek, M. F., Dabek, F. & Balakrishnan, H. [2003]. Chord: a scalable peer-to-peer lookup protocol for internet applications, *IEEE/ACM Trans. Netw.* 11(1): 17–32.

[68] Tanenbaum, A. S. [2002]. *Computer Networks (International Edition)*, fourth edn, Prentice Hall.

[69] Tolone, W., Ahn, G.-J., Pai, T. & Hong, S.-P. [2005]. Access control in collaborative systems, *ACM Comput. Surv.* 37(1): 29–41.

[70] van Renesse, R., Minsky, Y. & Hayden, M. [1998]. A gossip-style failure detection service, *Proceedings of the IFIP International Conference on Distributed Systems Platforms and Open Distributed Processing*, Middleware '98, Springer-Verlag, London, UK, pp. 55–70.

[71] Vaquero, L. M., Rodero-Merino, L., Caceres, J. & Lindner, M. [2008]. A break in the clouds: towards a cloud definition, *SIGCOMM Comput. Commun. Rev.* 39: 50–55.

[72] Wall, D., Kudtarkar, P., Fusaro, V., Pivovarov, R., Patil, P. & Tonellato, P. [2010]. Cloud computing for comparative genomics, *BMC Bioinformatics* 11(1): 259.

[73] Wang, J., Varman, P. & Xie, C. [2011]. Optimizing storage performance in public cloud platforms, *Journal of Zhejiang University-Science C* 12(12): 951–964.

[74] Wu, L. & Buyya, R. [2010]. Service level agreement (sla) in utility computing systems, *CoRR* abs/1010.2881.

[75] Zhang, L., Gu, S., Wan, B., Liu, Y. & Azuaje, F. [2011]. Gene set analysis in the cloud, *Bioinformatics* 13: 1–10.

Chapter 6

Hardware Accelerated Molecular Docking: A Survey

Imre Pechan and Béla Fehér

Additional information is available at the end of the chapter

http://dx.doi.org/10.5772/48125

1. Introduction

Hardware acceleration is the general concept of applying a specialized hardware for a given problem instead of an ordinary CPU in order to get lower processing time. General purpose CPUs can be considered as a totally general platform suitable for executing virtually any software or algorithm. Application specific accelerators have a custom architecture that fits the needs of a certain family of algorithms. As a consequence, they are able to outperform CPUs by orders of magnitude in a special application area but they are unfit for other, more general tasks. In contrast to normal CPUs, which are essentially serial machines executing instructions sequentially, hardware accelerators use parallel architectures which allow them to exploit the parallelism available in the given application by performing independent operations simultaneously.

The most important examples of hardware accelerators are graphics processing units (GPUs) and field-programmable gate array devices (FPGAs). GPUs are special many-core processors optimized for 3D rendering and image processing purposes. GPU devices are nowadays part of any desktop PC configurations and they can be programmed with general purpose programming languages. These facts make them an easily accessible and cost-effective accelerator platform and explain why they are used more and more frequently even in applications that are not graphics-related (general purpose GPU programming). FPGAs are programmable logic devices consisting of hundreds of thousands of general logic elements whose interconnection can be configured by the user. Thus FPGAs have a highly flexible architecture that allows to implement a totally custom digital hardware without the enormous cost of designing and manufacturing an application-specific integrated circuit (ASIC). When using an FPGA as a hardware accelerator a custom logic device is realized in the FPGA whose only purpose is to execute the algorithm to be accelerated as effectively as possible; thus the algorithm is usually implemented as pure hardware instead of software.

Hardware accelerators such as GPUs or FPGAs are utilized in many scientific applications, when the time-consuming operations make it impractical or even impossible to use ordinary CPUs. Bioinformatics is not an exception; it includes many problems and algorithms which are computationally expensive due to the large amount of data to be processed or the complex operations involved. Typical examples are different sequence alignment algorithms, protein structure prediction algorithms and molecular dynamics simulations which were implemented on various accelerator platforms several times.

Molecular docking is another key field of bioinformatics whose purpose is to determine the binding geometry of molecules and is used by the pharmaceutical industry for identifying drug candidate compounds. Docking algorithms are usually computationally demanding since they consist of generating and evaluating a large amount of different molecule conformations and placements. However, these different placements can often be processed simultaneously and evaluating a single placement usually offers further parallelization possibilities. These facts make molecular docking an ideal target for hardware acceleration. In accordance with this, several GPU- and FPGA-based docking implementations were reported applying different approaches for hardware acceleration. In this chapter our purpose is to give a general overview of the most interesting implementations and to compare them with respect to the applied parallelization, applicability and achieved speedup. The remainder of this chapter is organized as follows. Section 2 surveys the concept and methods of molecular docking. Section 3 gives a general overview of FPGA and GPU devices. Section 4 and 5 introduce the existing FPGA- and GPU-based docking implementations, respectively. Finally, Section 6 surveys the current state and perspectives of hardware accelerated molecular docking.

2. Overview of molecular docking

Molecular docking is a computer simulation technique for determining the possible binding position and binding energy of molecules whose initial 3D spatial structure is known. Many docking methods and software exist, which may be different in several respects such as the size and number of molecules involved, the applied docking algorithm, the applied chemical model or the modeling of molecular flexibility.

Molecular docking usually refers to docking a molecule to another one, that is, to determine the binding pose of the former relative to the latter. In case of protein-protein docking both of the molecules are large macromolecules. The more typical case is the protein-ligand docking when one of them is a small ligand molecule whose binding pose needs to be determined within the active site of a receptor. Since the computational complexity (the number of atoms, the size of the search space, etc.) differ by orders of magnitude, protein-protein and protein-ligand docking usually require different approaches. Although the number of molecules involved in the docking problem is generally two, some protein-ligand docking software allow to dock more than one ligand to a macromolecule simultaneously. For some software a good starting position has to be provided manually which is then refined by the algorithm; other ones are totally automated and try to find the docked position without any a priori knowledge.

Another important aspect is how the docking algorithm takes into account molecular flexibility. Rigid-body docking methods keep the structure of the molecules rigid, flexible algorithms consider one or both of the molecules flexible allowing their conformation to change. The two approaches correspond to the lock-key and the induced fit model, respectively. Rigid docking methods are usually much faster but may easily fail to find the proper binding position in case of molecules that actually undergo a conformational change upon binding. The most obvious way to model flexibility is to consider some bonds rotatable by allowing their torsional angle to change during docking. This method is effective in case of small ligands, but greatly increases the number of degrees of freedom and the computational complexity of the docking problem when applied for a large protein. As a consequence, protein flexibility is often taken into account only partially (allowing a few bonds of some side chains to rotate) or is modeled differently. One example is the soft receptor technique which allows small atomic collisions between neighboring protein and ligand atoms by reducing the repulsion energy term. The method is based on the assumption that the highly flexible protein could avoid the collision in practice by a low energy conformational change. Modeling flexibility in this way is computationally economic but may easily lead to invalid docked positions. Another straightforward technique is to keep the protein structure rigid and repeat the docking process with different pre-generated (or experimentally determined) protein conformations. Ultimately, this enables taking into account both protein and ligand flexibility even in case of rigid-body docking methods. The approach is also useful for considering the flexibility of rings within the ligand, which cannot be modeled with rotatable bonds; instead, a set of pre-generated, valid substructure conformations can be used during docking.

Although there are numerous different molecular docking algorithms, essentially each of them consists of two important components: a scoring function and a search method. The scoring function represents a chemical model and usually estimates the free energy of a geometrical arrangement of the molecules, thus it scores the given placement. The search method tries to find the ideal arrangement by sampling the search space according to a strategy. Docking can be viewed as an optimization problem where the global optimum of the scoring function is to be identified and the degrees of freedom are the variables describing the position, orientation and conformation of the molecules. Some docking methods apply one of the standard force fields as scoring function such as AMBER or CHARMM [1-4]. Other ones use empirical scoring functions that consist of a sum of terms representing different interaction types between the molecules; the term types are weighted with values determined empirically from a set of protein-ligand complexes [5, 6]. Knowledge-based functions are also typical which are derived from the statistical analysis of a large database containing molecular structures [7, 8]. The search methods applied by the different docking methods are also very diverse. One example is incremental reconstruction applied by the docking tools DOCK [4] and FlexX [9], which split the ligand to be docked and place the fragments one-by-one at the binding site. AutoDock [6] and GOLD [10] use genetic algorithms as global optimization methods. AutoDock Vina [11] applies a quasi-Newton BFSG algorithm along with Monte Carlo simulation. Other standard algorithms such as simulated annealing, tabu search or particle swarm optimization techniques are also common. A good overview of the general terms and concepts of molecular docking can be found in references [12-14].

The most important application area of molecular docking is computer-aided, structure-based drug design. Docking can be used for identifying drug candidates (potential inhibitors) for a given target receptor molecule. During virtual screening the members of a large ligand database are docked one by one to the target; promising compounds are subjected to further experiments. Virtual screening is extremely time-consuming; accelerating it can make the drug design process more effective. Trivially, this can be done by executing the docking runs of different molecules in parallel utilizing a lot of CPU cores. The other method is to accelerate the applied docking algorithm itself, potentially by an FPGA- or GPU-based hardware accelerator.

3. Accelerator platforms

3.1. FPGA devices

A field-programmable gate array is a programmable logic device - an integrated circuit with a flexible hardware architecture that can be configured to implement a specific functionality. FPGAs represent a trade-off between highly flexible, general purpose microprocessors and high-performance application-specific integrated circuits (ASICs). FPGA devices execute the required computation with a specific hardware architecture just like ASICs. Although they are not as efficient in terms of performance and power consumption, implementing a custom hardware in a 100-1000$ FPGA does not require to manufacture a new chip which is affordable only in case of large-scale production. In addition, FPGAs can be reconfigured many times. Thus they can be considered general-purpose similarly to CPUs but due to the applied custom architecture they can be orders of magnitude faster in case of a specific application.

The two major FPGA vendors, Xilinx and Altera offer a wide range of FPGAs and FPGA families with different capabilities, the performance and complexity of the devices is also continuously growing; however, the basic architecture remains the same. FPGA devices consist of a large number of similar basic logic blocks or cells arranged usually in rows and columns on the chip and a configurable interconnect structure. Figure 1. shows a simplified diagram of the basic logic block (slice) of a Xilinx Virtex-4 FPGA. The slice consists of two 4-input LUTs (look-up tables), two D flip-flops, carry logic supporting chaining of neighboring slices for high-performance arithmetic operations and routing resource configurable by multiplexers. A 4-input look-up table is a simple $2^4=16$ bit memory element that can realize any four-variable logic functions when initialized with the truth table of the corresponding function. D flip-flops are 1-bit registers that capture and store the value of the D input at every active CLK clock edge. Thus LUTs are the basic resources of the FPGA for implementing combinational logic and D flip-flops for sequential logic, respectively. In addition to the general logic resources FPGAs usually include special purpose cells such as dedicated memory blocks or DSP (digital signal processing) blocks consisting of adders and multipliers for arithmetic-intensive applications. FPGA-based accelerator cards are usually equipped with high-capacity external memory modules and high-speed interfaces like PCIe in addition to the FPGA.

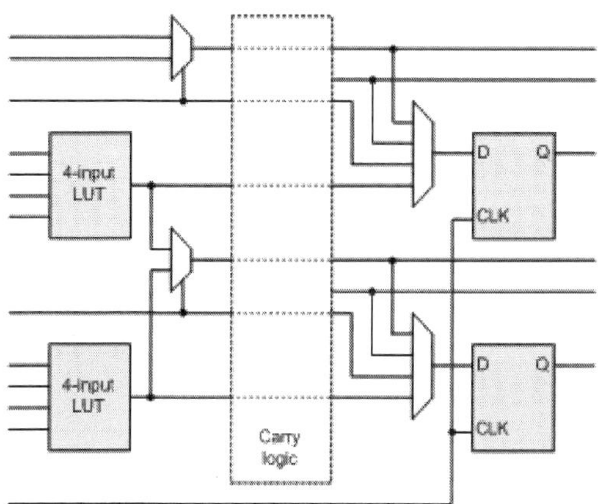

Figure 1. Virtex-4 slice

FPGA devices have an inherently parallel architecture which makes them suitable for high-performance computing applications. Different parts of an algorithm are executed by different hardware elements or modules; the execution can be simultaneous if the operations are independent. In data-parallel applications, where the same steps need to be performed on different data elements, the data can be distributed among many identical processing elements in the FPGA. In this case the achievable parallelism is limited only by the capacity of the device and the speed of the interface providing the input data. Another typical design concept is to apply a pipeline consisting of serially connected stages, which execute different steps of the same algorithm on different independent data elements.

Implementing an algorithm in an FPGA instead of a CPU may lead to a much shorter execution time; however, it usually requires more programming time and effort. The FPGA configuration can be defined with hardware description languages (HDL) such as VHDL and Verilog. HDLs allow the designer to describe the operation and interconnection of general digital circuits at a relatively high level (called register-transfer level). The HDL description is then mapped to the FPGA architecture by automatic tools. Further information regarding FPGA architectures, programming languages and design methodologies can be found in references [15-16].

3.2. GPU devices

Graphics processing units are massively parallel processors consisting of hundreds of processing cores, thus capable of executing hundreds of threads in parallel. Their architecture is optimized for data-parallel applications, which consist of instructions that have to be carried out on many different data elements. GPU operation is akin to the SIMD (single instruction multiple data) behavior – the parallel threads execute the same code but process independent input data. There are two main GPU manufacturers, AMD and NVIDIA, and although there are differences between the GPU architectures, the basic concepts are very similar. The same is true for the two widely used programming languages, CUDA and OpenCL. The former is developed by NVIDIA and is applicable to NVDIA devices only. OpenCL in turn is a standard parallel programming language supporting not only both GPU architectures but also multicore CPUs and heterogeneous platforms in general. The remainder of this section gives an overview of NVIDIA GPUs and CUDA since this is used by the majority of the GPU-based molecular docking implementations introduced in Section 5. However, the basic methodology and design patterns are very similar in case of OpenCL, only the terminology differs.

CUDA (Compute Unified Device Architecture) is the computing architecture of NVIDIA GPUs, which defines a parallel programming model based on high-level programming languages. CUDA C gives minimal extensions to the standard C language and provides an API, which enable the user to write a CUDA program consisting of serial code and special parallel functions called kernels. The former runs on the host CPU, the latter are executed K-times parallel by K different CUDA threads on the GPU. Threads of a kernel are grouped into thread blocks; the blocks in turn form a grid. Threads within the same block can communicate and synchronize with each other. This is not possible between different blocks of threads, since these are scheduled and executed in a random, non-deterministic order based on run-time decisions. This leads to automatic scalability; among ideal circumstances a GPU with twice as many processing cores can execute the same kernel twice faster.

The simplified hardware architecture can be seen on Figure 2. An NVIDIA GPU consists of multiprocessors. Each multiprocessor includes several processing cores, a large amount of registers, shared memory and a scheduler. In addition, each multiprocessor can access the external memory and has caches for texture and constant data access. When a kernel is launched, a certain number of thread blocks is assigned to every multiprocessor and becomes active. A multiprocessor executes its active blocks logically in parallel, and it manages, schedules and executes the threads of its active blocks in groups of 32 threads called warps. Warps are executed physically in parallel, that is, a multiprocessor is able to execute the same operation of every 32 thread within a warp simultaneously in one or a few clock cycle. However, if threads of a warp take different execution paths after a conditional branch statement, the different instructions get serialized, that is, they are executed sequentially (warp divergence).

Keeping the number of active blocks and warps high is important since this helps keeping every multiprocessor of the GPU busy as well as since the scheduler can hide the instruction and memory access latencies by switching between active warps. The maximal number of blocks that can be active on a multiprocessor is limited by the register and shared memory usage of the block since these resources are split among the active blocks. On the other hand, internal register and shared memory access is very fast. Threads can access their own registers in parallel; shared memory is divided into banks, and can be accessed also in parallel, as long as parallel threads access different memory banks. This suggests that data should be stored in registers and shared memory whenever possible. External memory access is much slower, but if threads of a warp read from or write to a contiguous memory space, the memory operations can be coalesced and executed as a single access, which can greatly increase the effective memory bandwidth. Constant data access is faster than ordinary memory read operations since it is cached. All of the aspects mentioned above have to be taken into account when choosing data storage areas, grid and block sizes. Further information about GPU architectures and programming can be found in references [17-19].

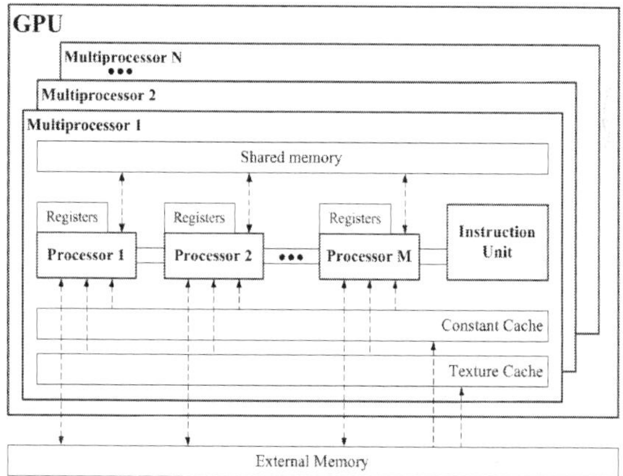

Figure 2. NVIDIA GPU architecture

4. Molecular docking on FPGA platforms

We believe that there are only three FPGA-based docking implementations which have been published until now. This chapter introduces all of them: a docking engine using 3D correlation, its successor, the FPGA-based implementation of the PIPER [20] docking program, and the FPGA-based acceleration of AutoDock.

4.1. Docking with 3D correlation

This implementation is described in references [21, 22]. The applied algorithm uses 3D correlation which is a common rigid-body docking technique. The molecules to be docked are represented with 3D grids whose voxels consist of pre-calculated values expressing some property of the molecule at the corresponding spatial location related to the binding affinity. In order to evaluate an arrangement the two grids are shifted relative to each other, then the voxels are multiplied pairwise and the values are summed to get the final score. By calculating the whole correlation array every possible translational position is evaluated.

This process has to be repeated for each orientation to be investigated, which requires the rotation of one of the grids periodically. In case of correlation-based docking methods the applied search method is essentially exhaustive search. Obviously the molecules are treated rigid during docking, since their structure is hard-coded in the grids.

The CPU-based docking programs using correlation usually replace it with Fourier transformation (FFT) and multiplication, which can be much faster on serial machines. The described FPGA-based implementation, however, performs direct correlation which can be effectively implemented with a highly parallel systolic chain in the FPGA. Another advantage of this method is that, by avoiding FFT, the operation for determining the voxel-voxel interaction is not restricted to multiplication; even non-linear functions can be used. In order to exploit this the implementation has a flexible structure; the design can be easily configured to adapt different scoring schemes. The initial implementation [21] used a very simple voxel type consisting of only two bits that distinguish molecule interiors from exteriors and mark the surface of the molecules. The final version allows using voxels with tuple data type that represent different effects including directional interactions like hydrogen bonding.

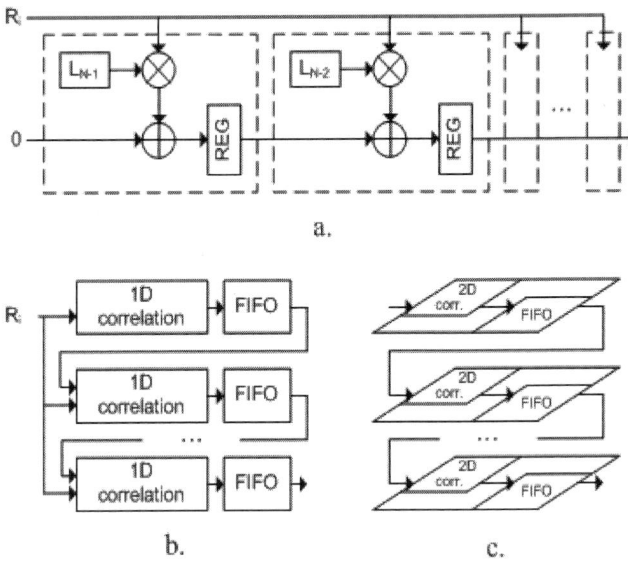

Figure 3. Systolic 3D array architecture [22, 23]

The core element of the implementation is a systolic 3D correlation array consisting of cells. Each cell stores one voxel of the grid corresponding to the smaller (ligand) molecule. The receptor grid is stored in external memory. Instead of rotating the ligand at the beginning of each new correlation cycle, rotated orientation is obtained by reading the receptor voxels in rotated order. Thus rotation is performed on the fly by the address logic and uses only little of the FPGA resources. The receptor voxels read from the memory are also rotated in case of directional data types; then they are passed to the systolic array. Figure 3/a shows a 1D correlation array consisting of pipelined cells. Each cell executes a pairwise operation (in this case, a multiplication) defined by the scoring method on its ligand voxel (L_i) and the receptor voxel (R_i) available at its input, then adds it to the sum received from the previous cell and stores the result in a register. 1D correlation arrays are connected by FIFO delay lines to form a 2D correlation plane; planes in turn are connected by delay planes to obtain the 3D correlation array (Figure 3/b and 3/c). Due to the pipeline-based structure, the systolic array produces the result of one position evaluation (correlation) in each clock cycle, and the achieved parallelism is proportional to the number of ligand grid voxels. Due to the large amount of output data the resulting grids are not sent to the host machine directly. Instead, a data reduction filter module detects the best score (local maximum) within each subblock of the correlation array; only these promising docked positions are returned for further analysis. Certain parts of the FPGA design are configurable according to the applied voxel word with and type, score data type, and the applied pairwise scoring operation in order to support various force laws and scoring schemes.

Performance tests were carried out with a Xilinx Virtex-II Pro XC2VP70-5 FPGA. Results were compared to a software running on a 3 GHz Xeon CPU. The software applied direct correlation since FFT proved to be slower for the applied small problem sizes. FPGA speedup varied between ×100-1000 according to the scoring method used.

4.2. PIPER on FPGA

The docking engine described in reference [23] is a modified, extended version of the docking core introduced in Section 4.1 and implements the PIPER software [20]. PIPER is based on 3D correlation and calculates it with the standard FFT method. The scoring function of PIPER consists of the weighted sum of different terms represented with separate grids; as a consequence, several independent forward FFTs have to be performed during evaluation. In addition to the van der Waals repulsion and attraction terms and the electrostatic interaction, desolvation effect is taken into account as well. The latter is described by a pairwise potential which is transformed to correlation grids with eigenvalue-eigenvector decomposition. Grids corresponding to low eigenvalues are often discarded which reduces computational complexity but retains the accuracy of the algorithm [20].

The original implementation described in Section 4.1 was modified in a variety of ways to support multiple energy grids as well as to allow docking of larger molecules which would not fit in the systolic 3D array, thus permitting even protein-protein docking. The basic cell element of the systolic array is extended to process the independent grids in parallel; as a consequence, each correlation is performed simultaneously. At the end of every 1D correlation array a new weighted scorer module sums the partial correlation results with respect to the weights defined in PIPER. New FIFOs are used to propagate the output of a scorer module to the input of the next one. Calculating the weighted sum at the end of every

1D array requires a balanced amount of multipliers and FIFOs (block RAMs) of the FPGA keeping the resource utilization optimal. To support large molecules both the receptor and the ligand are stored in external memories. The ligand grid is partitioned into subgrids small enough to fit the size of the correlation array. Correlation is performed piece-wise; each subgrid is first loaded to the FPGA, then the receptor voxels are streamed through the array.

The docking engine was implemented on an Altera Stratix-II EP2S180 and was validated against the original PIPER software. FPGA performance, however, was determined with post place-and-route simulations supposing an Altera Stratix-III EPSL340. Performance was compared to the original PIPER code, its multithreaded version, as well as a GPU-based implementation of PIPER (introduced in Section 5.1). The host CPU was a quad-core Intel Xeon 2 GHz CPU, the GPU code run on an NVIDIA Tesla C1060 device. The measured FPGA speedup depended greatly on the ligand grid size. In case of a 4^3 ligand grid speedup of the correlation task only and that of the whole application was almost ×1000 and ×37, respectively, compared to the single-core PIPER. However, it dropped exponentially with respect to the ligand size, decreasing below the ×16 speedup of the GPU at grid edge size 16 and below the ×3 speedup of the quad-core version at grid edge size 32. The reason for this is that the FFT method applied by the CPU and the GPU became greatly superior to direct convolution at this problem sizes.

4.3. AutoDock on FPGA

References [24, 25] introduce our own FPGA-based docking implementation, the acceleration of the AutoDock [6] docking software. AutoDock is applicable basically for protein-ligand docking and models molecular flexibility with rotatable bonds. AutoDock uses a semi-empirical scoring function that consists of weighted terms representing van der Waals and electrostatic interactions, hydrogen bonding and desolvation. The scoring function gives the energy contribution of one non-bonded atom pair; this value has to be summed over all movable atom pairs of the system to determine the score. To reduce computational complexity AutoDock represents the rigid part of the receptor molecule with pre-calculated potential grids. Thus the energy contribution of a given ligand atom and the whole receptor can be determined with trilinear interpolation and iterating over the receptor atoms is not necessary. AutoDock uses a standard genetic algorithm (GA) as search method. Genetic algorithms generate sets of potential solutions (generations of entities) iteratively. Solutions are represented with values of the degrees of freedom (called genes) and are created by combining the genes (crossover) of selected previous entities (selection) and altering them randomly (mutation). In addition to the genetic algorithm AutoDock subjects some selected entities of each generation to an iterative local search method (LS) similar to hill climbing, which greatly increases the effectiveness of the algorithm.

AutoDock was implemented on the SGI RASC RC100 module on a Xilinx Virtex-4 LX200 FPGA. The design consists of four main blocks organized as a three stage pipeline (Figure 4). The first pipeline stage executes the genetic algorithm, that is, it generates the genes of a new entity periodically. The second stage calculates the positions of the atoms of the ligand based on the input gene values. This step consists mainly of performing atomic rotations according to the positions of rotatable bonds and to the orientation of the ligand. The third stage includes two modules. One of them determines the receptor-ligand interaction energy based on the potential grids stored in external memory, that is, it performs a trilinear interpolation for each ligand atom; the other one calculates the energy contribution of each movable ligand atom pair by evaluating the scoring function directly. Each of the four modules consists of massively parallel, fine-grained internal pipelines; as a consequence, all of them are able to produce a new result of the realized operation in each clock cycle. The first module generates a new gene value, the second one performs the rotation of an atom, the other ones calculate the interaction energy of a ligand atom and the receptor molecule or that of an internal ligand atom pair in every clock cycle.

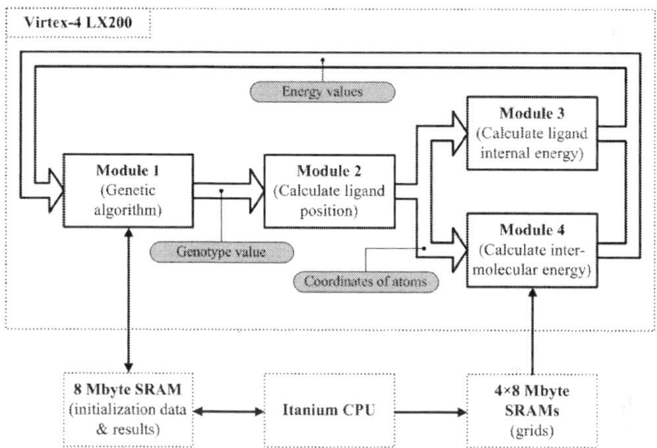

Figure 4. FPGA core implementing AutoDock [24]

In order to increase the performance of the docking engine the implemented algorithm slightly differs from the original AutoDock code and uses fixed-point arithmetic that fits better the FPGA architecture. According to test runs these differences does not degrade the accuracy of docking. Performance tests showed that the FPGA-based implementation yields an average speedup of ×23 over AutoDock running on a 3.2 GHz Intel Xeon CPU; the actual speedup varied between ×10-40 according to the structure and size of the molecules.

5. GPU-based implementations

Compared to the relatively small number of FPGA-accelerated docking engines, quite a lot of GPU-based solutions have been reported, which clearly indicates the advantages of GPUs over FPGAs in terms of accessibility and programming effort. It is neither reasonable nor possible to introduce every one of them. Instead, we aim at describing a wide variety of different approaches, and we tried to select the most promising implementations. Two of the docking codes introduced in the following subsections were implemented also on FPGA.

5.1. PIPER on GPU

The authors of the FPGA-based PIPER (Section 4.2) published also a GPU-based version [26]. In case of the FPGA the FFT applied in PIPER was replaced with direct correlation, which can be executed by a very effective standard structure in the FPGA. On GPU, both FFT and direct correlation were implemented and they proved to be advantageous at different ligand grid sizes. Other steps such as summing the grids and filtering the results by identifying local maxima also run on the GPU; although they comprise only a few percent of total PIPER runtime, executing them on the CPU would have limited the achievable speedup. The only exception is re-calculation of the ligand grid according to the current orientation and charges, which run on the host CPU.

3D correlation includes a lot of parallelism which can be exploited on a GPU as easily as on FPGA. Each voxel of the result grid can be calculated by a different thread simultaneously; in addition, correlation of grids representing different terms can be performed in parallel. In this implementation two different approaches are applied whose performance turned out to be similar: assigning each 2D plain to a different tread block, and assigning the same part of each 2D plain to a thread block. Receptor grid is stored in the external memory of the GPU due to its size and since it has to be available for each thread block (multiprocessor). Ligand grid is stored in shared or constant memory if possible; if the grid is small enough grids corresponding to multiple ligand orientations are stored and processed in parallel, which leads to further performance improvement.

Forward and inverse FFT is executed with the standard NVIDIA CUFFT library consisting of optimized FFT-related CUDA functions. Receptor grids are calculated by the CPU, moved to the GPU memory and transformed by the GPU only once at initialization. Ligand grids are re-calculated, copied and transformed for each ligand orientation. Voxels of the transformed receptor and ligand grids are multiplied pairwise by the GPU. The CUDA implementation is trivial, since each voxel pair can be processed independently by a different thread. Finally the product grid is inverse transformed.

The final step of each orientation evaluation is to sum the result grids according to the PIPER coefficients and find the voxels with the best scores corresponding to the best translational poses. PIPER uses several different sets of weights; these are assigned to different thread blocks in the GPU. Each block performs averaging according to the given set of weights. Individual threads process different parts of the grid. During averaging each thread identifies the best score of the grid part assigned to it and stores it in shared memory. Finally a single thread iterates over the scores to find the best one. Clearly the last filtering step could be implemented on the GPU the less effectively; if the number of coefficient sets is less than that of the multiprocessors, certain processors are not utilized, and serial steps such as finding the very last best score leads to idle threads. The majority of the algorithm, however, suits well the GPU architecture.

Performance tests were carried out on the platforms already mentioned in Section 4.2; the CUDA code run on a Tesla C1060 GPU and was compared to the FPGA-based version and to PIPER running on a single core and on all the four cores of a 2 GHz Intel Xeon CPU. Speedup of the correlation task was about ×300 compared to the single core version at a minimal ligand grid size of 4, but decreased exponentially with respect to ligand size similarly to the FPGA-based implementation. FPGA speedup was about ×1000 in case of a 4^3 ligand grid, so in case of direct correlation the FPGA outperformed the GPU. The FFT-based GPU code achieved a speedup of about ×30 regardless the ligand size and proved to be faster than GPU-based direct correlation above ligand grid size 8^3. Worst-case speedup of the whole GPU application was ×17.7 and ×6.1 versus single core and quad-core PIPER, respectively, and was faster than the FPGA accelerated version if ligand grid size was above 8^3.

5.2. A general FFT-based approach

In reference [27] another CUDA implementation is presented that applies FFT for performing the correlation-based rigid docking algorithm. The approach is very similar to the one described in Section 5.1. The scoring function is very simple; it consists of two terms which represent the shape of the molecules and the electrostatic field. These terms are calculated over the 3D grid for the receptor and for each orientation of the ligand. Again, FFT is executed with the CUDA library.

The test environment consisted of a dual-core AthlonX2 3600+ CPU and an NVIDIA GeForce9800GT GPU. The GPU speedup proved to be about ×3-4, depending on the grid size and the angle step size between different ligand orientations. That is, for the same search space size a finer discretization of the grids (meaning higher number of grid voxels) and a finer discretization of the ligand orientation (leading to more different orientations to be evaluated) resulted higher speedup. The reason is that in this case the FFT-grid multiplication-IFFT steps became more dominant compared to the whole GPU algorithm, and these can be executed the most effectively on the GPU.

The achieved GPU performance seems to be lower with respect to the GPU-based PIPER (Section 5.1). Although the applied algorithms and implementation methods are similar, the achieved speedups are hard to compare due to the different hardware platforms. The GeForce 9800 GT includes about half the number of multiprocessors than Tesla C1060. CPU frequencies are the same but the architectures are very different; the applied AMD CPU is older than the Intel used in case of PIPER. The other possible explanation of the different performance improvements is that in case of PIPER several grids has to be processed during docking, which leads to more parallelism and requires more FFT computation; thus the advantages of the GPU can be exploited more effectively.

5.3. AutoDock on GPU

AutoDock is one of the best-known docking software; it was the most cited docking program in the ISI Web of Science database in 2005 [28]. This explains why it is a popular subject for GPU-based acceleration. There is even a related SourceForge project called gpuautodock. The following subsections focus on three different AutoDock implementations; each of them maps different parts of the original algorithm to the GPU architecture.

5.3.1. Acceleration based on profiling

This AutoDock implementation is described in a case study [29]. The authors followed a traditional way – they profiled the original code in order to identify the most time-consuming functions and ported only these to GPU. Two functions were selected – eintcal() and trilininterp() – which together accounted for about 63% of the total runtime. The former calculates the internal energy of the ligand molecule, that is, it evaluates the scoring function for each ligand atom pair whose distance can change due to rotatable bonds. The latter is called for each ligand atom during the calculation of receptor-ligand intermolecular energy to perform interpolation based on the pre-calculated potential grids.

Each time these functions are called the corresponding CUDA kernel is executed instead of the original function. In both cases the number of threads within the kernel equals to the number of ligand atoms. This molecule usually consists of a few tens of atoms, which is a very low number compared to the GPU capabilities leading to a poor GPU utilization ratio. In addition, before each kernel call some data is transferred from the main memory to the GPU according to the current ligand position; these frequent memory transfer operations further decrease the performance.

According to test runs, which were executed on an NVIDIA GeForce GTX 280 GPU, the GPU accelerated application could not achieve speedup but was slower than the CPU for typical ligand sizes. Performance improvement was obtained only if the number of atoms (threads) was in the range of 10^4, which is not a realistic use case. The reasons are mentioned above. Accelerating only a few computationally expensive functions without restructuring the original code is straightforward and does not require much programming effort; however, it does not allow to exploit all the parallelism available in the algorithm, and also limits the maximal achievable speedup according to Amdahl's law.

5.3.2. Acceleration excluding local search

AutoDock includes further parallelism that is not exploited by the implementation described in Section 5.3.1. It uses a genetic algorithm as search method, which can be parallelized easily; each entity of the next generation can be created and evaluated simultaneously by different processing cores. The default population size is 150, which makes this approach promising with respect to GPU-based acceleration. However, AutoDock also applies an iterative local search method in addition to the GA, which is executed only on a few percent of the population (6%, that is, averagely 9 entities by default). Executing local search of different entities in parallel is possible, but would lead to low GPU utilization. In addition, performing the local search algorithm on an entity may consist of hundreds of iterations (energy evaluations); that is, executing the whole LS on CPU would greatly reduce the achievable speedup.

To overcome this problem, the authors of reference [30] chose to exclude the local search from the algorithm, but implement virtually every other part of AutoDock (the genetic algorithm, the ligand position calculation and the scoring function evaluation) on the GPU. Although the genetic algorithm (generating the degrees of freedom according to the GA rules) is usually not time-critical, leaving it on the host CPU would require periodic CPU-GPU memory transfer operations, which is avoided if it is executed by the GPU.

The main idea behind the implementation is to assign a different thread block to each entity of the new generation, whose threads cooperatively execute the different steps on the given entity. Another scheme was also tried where different entities were assigned to different threads, but this leads to low GPU utilization in case of typical population sizes – using default size, the number of parallel threads would be only 150 instead of 150 multiplied by the thread block size. The coordinates of ligand atoms are stored in the fast shared memory, which is crucial since every step of the scoring function evaluation modifies or reads this data and each thread of the block has to access it. During evaluation, each thread block first determines the atom positions (using two kernels for calculating the ligand conformation and orientation). Independent rotations of different atoms can be executed by different threads of the block. Then each thread performs trilinear interpolation for a different ligand atom (determining the atom-receptor intermolecular energy), and each thread evaluates the scoring function directly for a different ligand-ligand atom pair. Trilinear interpolation offers a further optimization, since NVIDIA GPUs support the fast access of 3D data by hardware.

Parallelization of the GA operators (selection, crossover and mutation) is also straightforward. Selection requires to calculate the relative score (fitness) of the entities compared to the average score, which can be performed for each entity simultaneously. Genes of the new entities can be generated by crossover and mutation in parallel by threads of the block assigned to the entity.

The test platform included an AMD Athlon 2.4 GHz and an NVIDIA Tesla C1060. Validation of the CUDA code was performed by using the same random seeds and comparing the output to that of original AutoDock. The results differed slightly only due to the single precision arithmetic applied in the GPU. The speedup of the different kernels depended highly on the population size. At default size speedup of the scoring function evaluation proved to be ×50, the selection and crossover ×1.25 and ×2.75. In case of mutation no speedup was obtained. The overall speedup of the algorithm was ×10 for a population size 50, it increased to ×20 for the default size and become saturated at 10000 yielding a speedup of ×47 over the CPU. On one hand, the GA operators could be implemented on the GPU much less effectively than the fitness evaluation. The probable reason is that they are much more control-intensive than the different steps of the scoring function evaluation consisting of a lot of arithmetic operations. On the other hand, executing GA on the CPU would certainly decrease the speedup due to the additional transfer operations between the CPU and GPU memory.

5.3.3. Acceleration including local search

Implementing AutoDock on CUDA without local search as in Section 5.3.2 clearly offers a straightforward parallelization scheme that avoids GPU underutilization. However, the local search process usually increases docking accuracy of AutoDock significantly [31]. In order to include the LS in the implemented algorithm and simultaneously achieve a high speedup we ported AutoDock to CUDA exploiting a further high-level parallelization possibility [25]. Due to the heuristic nature of the search algorithm, often several (10-100) different docking runs are performed with AutoDock for the same receptor-ligand complex. This increases the reliability of the results as well as helps identifying multiple valid docked poses. Since these docking runs are totally independent from each other, they can be executed in parallel.

Our implementation includes two CUDA kernels. In each generational cycle, first Kernel A is launched that creates and evaluates a whole population; then Kernel B is launched for performing LS on the selected entities. The two kernels call the same CUDA functions for scoring function evaluation; they differ only in how the degrees of freedom are generated (using either GA or LS rules). Basically, our implementation is quite similar to the one introduced in Section 5.3.2. Each thread block of the kernels is assigned to a different entity. Threads within a thread block generate different gene values, calculate independent rotations, and process different ligand atoms or atom pairs during scoring function evaluation. However, in case of kernel A a thread block is launched for each new entity of every independent docking run; in case of kernel B a block is launched for each entity of every run which is selected for local search.

The advantage of this method is that local search is included which allows preserving docking accuracy, and even significant performance improvement can be achieved if the number of independent runs is high enough. The performance improvement, however, depends strongly on this number and in case of too few parallel runs the GPU is underutilized during LS, which leads to a low speedup.

Test runs were carried out on an NVIDIA GeForce GTX 260 GPU; performance was compared with that of AutoDock running on a 3.2 GHz Intel Xeon CPU. In case of only one docking run the GPU achieved a low, ×2-5 speedup depending on the ligand structure and size. In case of 10 and 100 independent runs the average speedup proved to be ×30 and ×65, respectively, for a large set of ligands.

Our FPGA-based AutoDock implementation described in Section 4.3 achieved an average speedup of ×23. This value does not depend on the number of docking runs since the FPGA executes only one at a time. Due to the applied three stage pipeline (Figure 4) only three entities are processed simultaneously in the FPGA. On the contrary, the GPU applies a brute force approach by processing each entity of every run in parallel. The low level (per rotation, per atom, etc.) parallelization possibilities are exploited by fine-grained pipelines in the FPGA very effectively; this allows the FPGA-based implementation to achieve a significant speedup regardless the number of runs. As a consequence, the FPGA is faster than the GPU for a low number of runs. Further advantage of the FPGA architecture is that implementing local search is not problematic. However, if the number of runs is high enough, the GPU outperforms the FPGA; that is, similarly to the FPGA and GPU-based PIPER (Section 4.2 and 5.1) the two platforms are advantageous at different parameter ranges.

5.4. MolDock on GPU

Reference [32] describes the GPU-based acceleration of the MolDock [33] docking software. MolDock is very similar to AutoDock: it models molecular flexibility with rotatable bonds, its scoring function consists of the summation of pairwise energy terms, it uses pre-calculated potential grids for representing the receptor during docking and it applies a genetic (evolutionary) algorithm as search method. Differences are the actual form of the energy terms (which is virtually irrelevant from the point of view of parallelization) and the lack of local search.

Due to the similar algorithms the basic implementation schemes are practically the same as the ones described in Section 5.3.2 and 5.3.3; the gene values and atoms of every entity are distributed among the threads and are processed in parallel. Although no local search process is used, independent docking runs are performed in parallel to increase GPU utilization ratio (like in Section 5.3.3). Due to these similarities the implementation is not described here in more details. However, we would like to emphasize an apparent difference regarding how different jobs are aligned to the threads of the kernel.

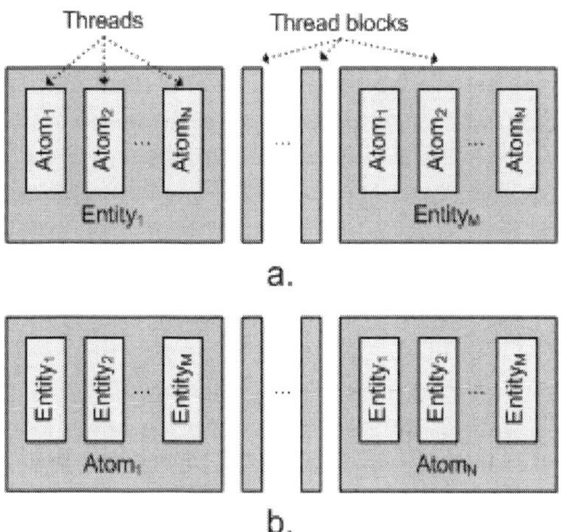

Figure 5. Job alignment comparison

intensive applications like the majority of the docking algorithms, although clearly there are problem domains where FPGAs remain superior.

Author details

Imre Pechan
evopro Informatics and Automation Ltd, Budapest, Hungary

Béla Fehér
Department of Measurement and Information Systems,
Budapest University of Technology and Economics, Budapest, Hungary

Acknowledgement

We would like to thank evopro Informatics and Automation Ltd for supporting our work and providing access to the necessary hardware and software tools.

7. References

[1] Brooks BR, Bruccoleri RE, Olafson BD, States DJ, Swaminathan S, Karplus M (1983) CHARMM: A Program for Macromolecular Energy, Minimization, and Dynamics Calculations. J. Comput. Chem. 4: 187-217.

[2] Cornell WD, Cieplak P, Bayly CI, Gould IR, Merz KM, Ferguson DM, Spellmeyer DC, Fox T, Caldwell JW, Kollman PA (1995) A Second Generation Force Field for the Simulation of Proteins, Nucleic Acids, and Organic Molecules. J. Am. Chem. Soc. 117: 5179-5197.

[3] Grosdidier A, Zoete V, Michielin O (2011) Fast docking using the CHARMM force field with EADock DSS. J. Comput. Chem. 32: 2149-2159.

[4] Ewing TJA, Makino S, Skillman AG, Kuntz ID (2001) DOCK 4.0: Search Strategies for Automated Molecular Docking of Flexible Molecule Databases. J. Comput. Aided Mol. Des. 15: 411-428.

[5] Friesner RA, Banks JL, Murphy RB, Halgren TA, Klicic JJ, Mainz DT, Repasky MP, Knoll EH, Shelley M, Perry JK, Shaw DE, Francis P, Shenkin PS (2004) Glide: A New Approach for Rapid, Accurate Docking and Scoring. 1. Method and Assessment of Docking Accuracy. J. Med. Chem. 47: 1739-1749.

[6] Huey R, Morris GM, Olson AJ, Goodsell DS (2007) A Semiempirical Free Energy Force Field With Charge-Based Desolvation. J. Comput. Chem. 28: 1145-1152.

[7] Muegge I, Martin YC (1999) A General and Fast Scoring Function for Protein-Ligand Interactions: A Simplified Potential Approach. J. Med. Chem. 42: 791-804.

[8] Gohlke H, Hendlich M, Klebe G (2000) Knowledge-Based Scoring Function to Predict Protein-Ligand Interactions. J. Mol. Biol. 295: 337-356.

[9] Rarey M, Kramer B, Lengauer T (1997) Multiple Automatic Base Selection: Protein–Ligand Docking Based on Incremental Construction Without Manual Intervention. J. Comput. Aided Mol. Des. 11: 369-384.

[10] Jones G, Willett P, Glen RC, Leach AR, Taylor R (1997) Development and Validation of a Genetic Algorithm for Flexible Docking. J. Mol. Biol. 267: 727-748.

[11] Trott O, Olson AJ (2010) AutoDock Vina: Improving the Speed and Accuracy of Docking with a New Scoring Function, Efficient Optimization, and Multithreading. J. Comput. Chem. 31: 455-461.

[12] Teodoro ML, Phillips GN, Kavraki LE (2001) Molecular Docking: A Problem With Thousands of Degrees of Freedom. IEEE Int. Conf. on Robotics and Automation, 2001 May 21-26, Seoul, Korea.

[13] Dias R, de Azevedo WF (2008) Molecular Docking Algorithms. Curr. Drug Targets 9: 1040-1047.

[14] Kavraki LE (2007) Protein-Ligand Docking, Including Flexible Receptor-Flexible Ligand Docking. Receptor 1-19. Available: http://cnx.org/content/m11456/latest/. Accessed 2012. 04. 29.

[15] Hauck S, DeHon A (2007) Reconfigurable Computing - The Theory and Practice of FPGA-Based Computation. Morgan Kaufmann. 944 p.

[16] Kilts S (2007) Advanced FPGA Design - Architecture, Implementation, and Optimization. Wiley. 352 p.

[17] NVIDIA CUDA C Programming Guide. Available: http://developer.nvidia.com/nvidia-gpu-computing-documentation. Accessed 2012. 04. 29.

[18] Sanders J, Kandrot E (2010) Cuda by Example - An Introduction to General-Purpose GPU Programming. Addison-Wesley. 312 p.

[19] AMD Accelerated Parallel Processing OpenCL Programming Guide. Available: http://developer.amd.com/sdks/AMDAPPSDK/documentation/Pages/default.aspx. Accessed 2012. 04. 29.

[20] Kozakov D, Brenke R, Comeau SR, Vajda S (2006) PIPER: An FFT-Based Protein Docking Program with Pairwise Potentials. Proteins 65: 392-406.

[21] VanCourt T, Gu Y, Herbordt MC (2004) FPGA Acceleration of Rigid Molecule Interactions. 12th Ann. IEEE Symp. on Field-Programmable Custom Computing Machines, 2004 Apr. 20-23, Napa, USA.

[22] VanCourt T, Gu Y, Mundada V, Herbordt MC (2006) Rigid Molecule Docking: FPGA Reconfiguration for Alternative Force Laws. EURASIP J. on Applied Signal Processing 2006: 1-10.

[23] Sukhwani B, Herbordt MC (2010) FPGA Acceleration of Rigid-Molecule Docking Codes. IET Comput. Digit. Tech. 4: 184-195.

[24] Pechan I, Fehér B, Bérces A (2010) FPGA-Based Acceleration of the AutoDock Molecular Docking Software. Conf. on Ph.D. Research in Microelectronics and Electronics, 2010 July 18-20, Berlin, Germany.

[25] Pechan I, Fehér B (2011) Molecular Docking on FPGA and GPU Platforms. Int. Conf. on Field Programmable Logic and Applications, 2011 Sept. 5-7, Chania, Greece.

[26] Sukhwani B, Herbordt MC (2009) GPU Acceleration of a Production Molecular Docking Code. 2nd Workshop on General Purpose Processing on Graphics Processing Units, 2009 Mar. 8, Washington, USA.

[27] Feng Z, Tian X, Chang S (2010) A Parallel Molecular Docking Approach Based on Graphic Processing Unit. 4th Int. Conf. on Bioinformatics and Biomedical Engineering, 2010 June 18-20, Chengdu, China.

[28] Sousa SF, Fernandes PA, Ramos MJ (2006) Protein-Ligand Docking: Current Status and Future Challanges. Proteins 65: 15-26.

[29] Micevski D, Kuiper M (2009) Optimizing Autodock with CUDA. VPAC Case Study. Available: http://www.vpac.org/?q=node/290. Accessed 2012. 04. 29.

[30] Kannan S, Ganji R (2010) Porting Autodock to CUDA. IEEE Cong. on Evolutionary Computation, 2010 July 18-23, Barcelona, Spain.

[31] Morris GM, Goodsell DS, Halliday RS, Huey R, Hart WE, Belew RK, Olson AJ (1998) Automated Docking Using a Lamarckian Genetic Algorithm and an Empirical Binding Free Energy Function. J. Comput. Chem. 19: 1639-1662.

[32] Simonsen M, Pedersen CNS, Christensen MH, Thomsen R (2011) GPU-Accelerated High-Accuracy Molecular Docking Using Guided Differential Evolution. 13th Ann. Conf. on Genetic and Evolutionary Computation, 2011 July 12-16, Dublin, Ireland.

[33] Thomsen R, Christensen MH (2006) MolDock: A New Technique for High-Accuracy Molecular Docking. J. Med. Chem. 49: 3315-3321.

[34] Korb O, Stützle T, Exner TE (2007) An Ant Colony Optimization Approach to Flexible Protein–Ligand Docking. Swarm Intell. 1: 115-134.

[35] Korb O, Stützle T, Exner TE (2011) Accelerating Molecular Docking Calculations Using Graphics Processing Units. J. Chem. Inf. Model. 51: 865-876.

[36] Ritchie DW, Venkatraman V (2010) Ultra-Fast FFT Protein Docking on Graphics Processors. Bioinformatics 26: 2398-2405.

[37] Ritchie DW, Kozakov D, Vajda S (2008) Accelerating and Focusing Protein-Protein Docking Correlations Using Multi-Dimensional Rotational FFT Generating Functions. Bioinformatics 24: 1865-1873.

[38] Guerrero GD, Sánchez HP, Wenzel W, Cecilia JM, García JM (2011) Effective Parallelization of Non-bonded Interactions Kernel for Virtual Screening on GPUs. 5th Int. Conf. on Practical Applications of Computational Biology & Bioinformatics, 2011 Apr. 6-8, Salamanca, Spain.

[39] Roh Y, Lee J, Park S, Kim J (2009) A Molecular Docking System Using CUDA. Int. Conf. on Hybrid Information Technology, 2009 Aug. 27-29, Daejeon, Korea.

Molecular Modeling

$$SD = \sqrt{\frac{\sum_{i=1}^{13} n_i \left(E_i - AM \right)^2}{\sum_{i=1}^{13} n_i}} \tag{3}$$

The harmonic means were calculated by first converting the binding energies into inhibition constant K_i

$$K_i = e^{\frac{1000 E_i}{RT}} \tag{4}$$

Where R is the Boltzmann constant and T is the temperature (298.15K). The harmonic means were calculated using below equation then converted back to HM binding energies.

$$\overline{K_i} = \frac{\sum_{i=1}^{13} n_i}{\sum_{i=1}^{13} \dfrac{n_i}{K_i(i)}} \tag{5}$$

2.4. Results

The obtained docking scores reveal that 6 compounds, specifically NSC211332, NSC141562, NSC109836, NSC350191, NSC1644640, and NSC5069, have higher binding affinity to the H1N1pdm neuraminidase protein than oseltamivir does.

Rank / Comp. ID	NSC	HM mean energy (kcal/mol)	Predicted Ki (μM)	AM mean energy (kcal/mol)	Standard deviation (kcal/mol)	Structure
1	211332	-12.05	0.001	-11.73	0.61	
2	141562	-10.14	0.037	-9.87	0.63	

Rank / Comp. ID	NSC	HM mean energy (kcal/mol)	Predicted Ki (µM)	AM mean energy (kcal/mol)	Standard deviation (kcal/mol)	Structure
3	109836	-9.74	0.072	-9.50	0.47	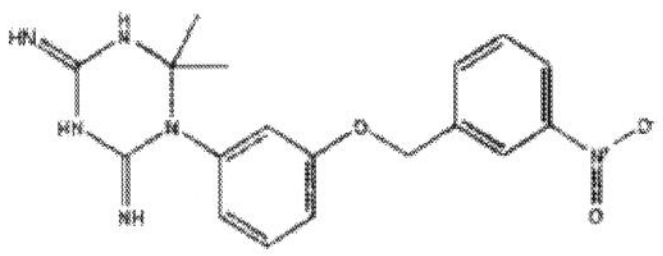
4	350191	-9.56	0.099	-8.77	0.68	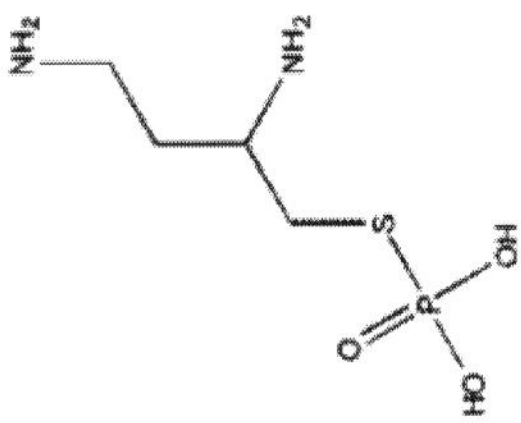
5	164640	-8.95	0.274	-8.78	0.42	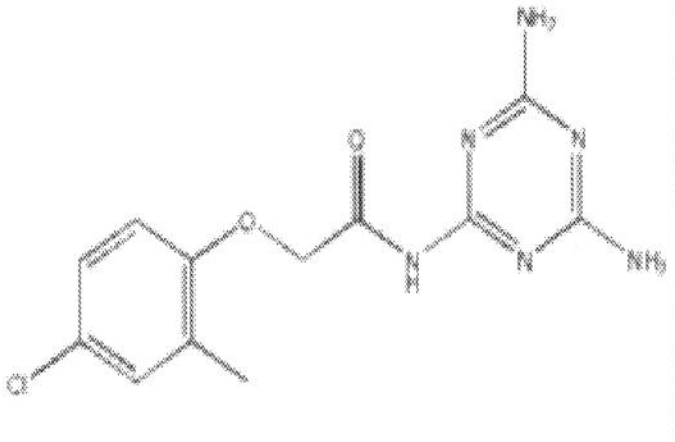
6	5069	-8.85	0.326	-8.23	1.11	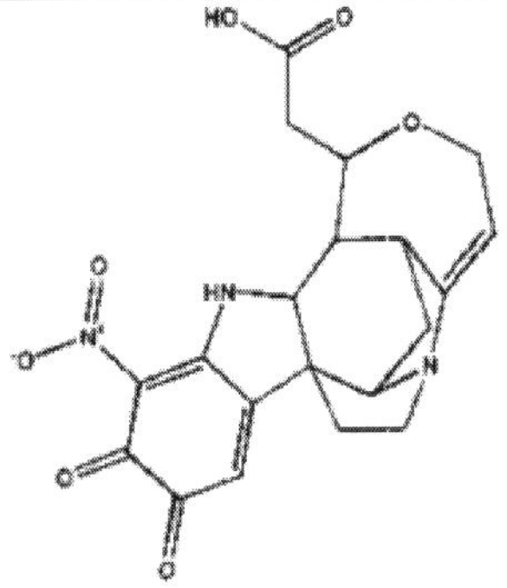

Rank / Comp. ID	NSC	HM mean energy (kcal/mol)	Predicted Ki (μM)	AM mean energy (kcal/mol)	Standard deviation (kcal/mol)	Structure
7	Oseltamivir	-8.69	0.429	-8.57	0.35	

Table 1. Docking results ranked by harmonic mean of binding energy. The first column is the final rank and also the compound ID. The predicted Ki calculated according to the harmonic mean binding free energies are also shown, as well as the arithmetic mean binding free energies and their corresponding standard deviations.

The results also confirm the observed effectiveness of oseltamivir against both the swine H1N1pdm and H5N1 flu. Detailed analysis on the hydrogen bond networks between top binding candidates with the swine H1N1pdm neuraminidase protein reveals that the mutations H274Y and N294S do not have any direct interactions with these compounds, and thus suggest the possibility of using the present top hit compounds for further computational and experimental studies to design new antiviral drugs against swine H1N1pdm flu virus and its variants (Hung et al., 2009). Furthermore, a more complete virtual screening using the full NCI diversity set (NCIDS) or larger sets from the ZINC database should be done to identify drug candidates that may have been missed in this study. The NCIDS has over 2000 compounds thus the full screening should be done with parallel automatic pipeline.

3. Investigation of drug resistant mechanism by MD simulation

Genetics study of influenza H5N1 virus isolated from patients who died despite being given oseltamivir showed that mutations H274Y or N294S confer high-level resistance to oseltamivir (De Jong et al., 2005 & Le et al., 2005)). There is even emerging evidence that these drug-resistant mutants pose the same risk with H1N1pdm, as shown by reported cases of the H274Y mutation of H1N1pdm (Guo et al, 2009). The rapid emergence of oseltamivir resistance in avian flu has already motivated numerous studies, both experimental and theoretical, to uncover the mechanisms of how point mutations in neuraminidase alter drug binding. Despite initial inroads, the current understanding of drug resistance remains incomplete and some conclusions are conflicting. For example, in one study, it was reported that the H274Y mutation disrupts the E276-R224 salt bridges, but in a separate study the same salt bridges were observed to be stable. We characterize the drug-protein interactions of oseltamivir bound forms of wild type avian H5N1 and swine H1N1pdm neuraminidases and how their mutations, H274Y and N294S, rupture these interactions to confer drug resistance through molecular modelling and atomistic MD simulation.

3.1. Computational details

The coordinates for the H5N1 neuraminidase bound with oseltamivir was taken from a monomer of the Protein Data Bank (PDB) structure 2HU4 (tetramer), while those of mutants H274Y and N294S were taken from structures 3CL0 (monomer) and 3CL2 (monomer) respectively. Even though the biological form of neuraminidase is tetrameric, its monomer contains a functionally complete active site and yields reasonable results in a prior study using MD simulations. The position for oseltamivir bound to H1N1pdm was adopted from its corresponding location in H5N1, as the two proteins' binding pockets differ only by residue 347 (which is Y in H5N1 and N in H1N1pdm), located on a loop at the periphery of the active site. Oseltamivir-mutant complexes of H1N1pdm were built by mutating H274Y and N294S of the H1N1pdm wild type model. In total, 6 systems were modelled and simulated for oseltamivir bound H5N1, and H1N1pdm wild type and H274Y and N294S mutants. Crystallographically resolved water molecules and a structurally relevant calcium ion near the native binding site for SA were retained and modelled in all simulated systems. The protein complexes were then solvated in a TIP3P water box and ionized by NaCl (0.152M) to mimic physiological conditions.

All simulations were performed using NAMD 2.7 and the CHARMM31 force field with CMAP correction. The ionized systems were minimized for 10,000 integration steps and equilibrated for 20 ns with 1 fs time steps. Following this, a 20 ns unconstrained equilibration production run was performed for subsequent trajectory analysis, with frames stored after each picosecond (every 1000 time steps). Constant temperature (T = 300 K) was enforced using Langevin dynamics with a damping coefficient of 1 ps−1. Constant pressure (p = 1 atm) was enforced using the Nosé-Hoover Langevin piston method. Van der Waals interaction cut-off distances were set at 12 Å (smooth switching function beginning at 10 Å) and long-range electrostatic forces were computed using the particle-mesh Ewald (PME) summation method.

Analysis included the calculation of an averaged electrostatic potential field over all frames of the trajectory using RMSD-aligned structures. Maps of the electrostatic potential field were calculated on a three-dimensional lattice. The long-range contributions to the electrostatics were approximated using the multilevel summation method (MSM), which uses nested interpolation of the smoothed pairwise interaction potential, with computational work that scales linearly with the size of the system. The calculation was performed using the molecular visualization program VMD that provides a graphics processing units (GPU) accelerated version of MSM to produce the electrostatic potential map. The GPU acceleration of MSM (Hardy et al., 2009) provided a significant speedup over conventional electrostatic summation methods such as the Adaptive Poisson Boltzman Solver (APBS), achieving a benchmark processing time of 0.2s per frame versus 180 seconds per frame (on a conventional CPU) using APBS69 for a 35,000 atom system, offering a speedup factor of about 900. The use of GPU acceleration enabled averaging the electrostatic potential field over all frames of the simulation trajectories. Hydrogen bond analysis utilized a distance and angle cut-offs of 3.5 Å and 60 degrees, respectively.

3.2. Results and discussion

Hydrogen bonds which form the primary interactions between oseltamivir and H5N1/H1N1pdm neuraminidases were shown in figure 4 and figure 5 accordingly. While the SA binding sites of H5N1 and H1N1pdm appear to differ mainly in the sequence of loop residue 347, it is not well understood whether antiviral drugs bind to each protein in the same manner, or if the drug resistant H274Y and N294S mutations disrupt critical hydrogen bonds. To address this question, the hydrogen bonds which form between oseltamivir and the residues lining the SA binding pockets of H5N1 and H1N1pdm were calculated for all simulation trajectories. R292 and R371 were observed to hydrogen bond with oseltamivir's carboxyl moiety, and E119 and D151 with oseltamivir's amino group (NH_3^+). The H274Y mutation, however, appeared to disrupt the hydrogen bonding of oseltamivir's acetyl group with R152, an interaction which was seen in the wild type and N294S systems for both simEQ1 and simEQ2. Prior analyses of crystallographic data alone suggested that Y347 forms a stable hydrogen bond with oseltamivir's carboxyl group and is the source of oseltamivir-resistance in the N294S mutant.

Our MD simulations however, reflect statistics collected from long timescale simulations which produce a dynamic picture of molecular interactions in greater detail and resolution than can be seen from a static crystal structure. In our simulations oseltamivir's carboxyl group primarily forms hydrogen bonds with R292 and R371, having little involvement with Y347. In fact, residue Y347 undergoes rotation to interact strongly with residue W295. Therefore, the speculation from previous studies, that the N294S mutation in the case of H5N1 actually destabilizes the hydrogen bonding between oseltamivir and Y347 to induce drug resistance, is not supported in our simulations.

The notable difference between H5N1 and H1N1pdm neuraminidases is the replacement of Y347 by N347 at the drug binding pocket. No conserved drug-protein hydrogen bond was observed for N347 in any of the three H1N1pdm simulations. Given the transient nature of even the N294S mutant induced hydrogen bond involving residue 347 in the case of H5N1, and the lack of interaction with residue 347 in any of the other simulations, it is highly unlikely that the single residue change (Y347 to N347), between the H5N1 and H1N1pdm strains significantly alters the drug-protein stability in regard to the hydrogen bond network involved. H274Y mutation induced disruption of the stable hydrophobic packing of oseltamivir's pentyl group in both H5N1 and H1N1pdm neuraminidases.

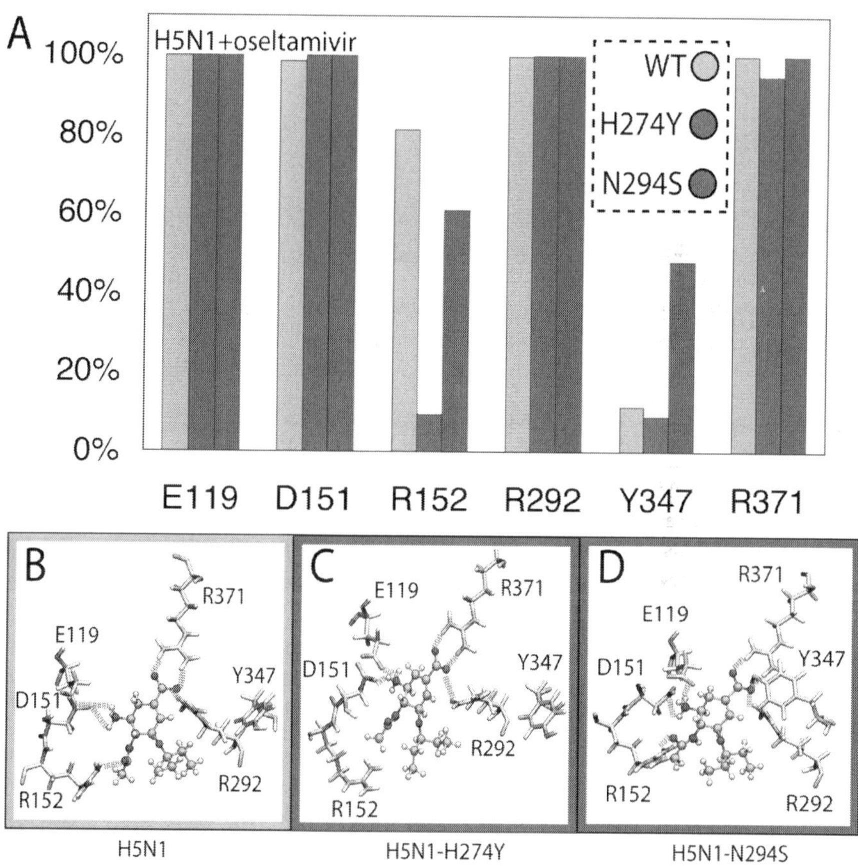

Figure 3. Network and occupancy of hydrogen bonds stabilizing oseltamivir in the SA binding pocket of wild type and drug-resistant mutant avian H5N1 neuraminidases, in simEQ1, simEQ3, and simEQ5.(A) shows histograms of the percent of hydrogen-bond occupancies for interactions between oseltamivir and residues E119, D151, R152, R292, Y347, and R371 across each simulation run. (B) through (D) are schematic views depicting the orientation of protein sidechains which form protein-drug hydrogen bonds.

Zoete V., Cuendet M. A., Grosdidier A., Michielin O., SwissParam, a Fast Force Field Generation Tool For Small Organic Molecules, J. Comput. Chem, 2011, in press. PMID: 21541964, DOI: 10.1002/jcc.21816.

Chapter 8

Using Molecular Modelling to Study Interactions Between Molecules with Biological Activity

María J. R. Yunta

Additional information is available at the end of the chapter

http://dx.doi.org/10.5772/54007

1. Introduction

To better understand the basis of the activity of any molecule with biological activity, it is important to know how this molecule interacts with its site of action, more specifically its conformational properties in solution and orientation for the interaction. Molecular recognition in biological systems relies on specific attractive and/or repulsive interactions between two partner molecules. This study seeks to identify such interactions between ligands and their host molecules, typically proteins, given their three-dimensional (3D) structures. Therefore, it is important to know about interaction geometries and approximate affinity contributions of attractive interactions. At the same time, it is necessary to be aware of the fact that molecular interactions behave in a highly non-additive fashion. The same interaction may account for different amounts of free energy in different contexts and any change in molecular structure might have multiple effects, so it is only reliable to compare similar structures. In fact, the multiple interactions present in a single two-molecule complex are a compromise between attractive and repulsive interactions. On the other hand, a molecular complex is not characterised by a single structure, as can be seen in crystal structures, but by an ensemble of structures. Furthermore, changes in the degree of freedom of both partners during an interaction have a large impact on binding free energy [Bissantz et al., 2010].

The availability of high-quality molecular graphics tools in the public domain is changing the way macromolecular structure is perceived by researchers, while computer modelling has emerged as a powerful tool for experimental and theoretical investigations. Visualisation of experimental data in a 3D, atomic-scale model can not only help to explain unexpected results but often raises new questions, thereby affecting future research. Models of sufficient quality can be set in motion in molecular dynamic (MD) simulations to move beyond a static picture and provide insight into the dynamics of important biological processes.

Computational methods have become increasingly important in a number of areas such as comparative or homology modelling, functional site location, characterisation of ligand-binding sites in proteins, docking of small molecules into protein binding sites, protein-protein docking, and molecular dynamic simulations [see for example Choe & Chang, 2002]. Current results yield information that is sometimes beyond experimental possibilities and can be used to guide and improve a vast array of experiments.

To apply computational methods in drug design, it is always necessary to remember that to be effective, a designed drug must discriminate successfully between the macromolecular target and alternative structures present in the organism. The last few years have witnessed the emergence of different computational tools aimed at understanding and modelling this process at the molecular level. Although still rudimentary, these methods are shaping a coherent approach to help in the design of molecules with high affinity and specificity, both in lead discovery and in lead optimisation. Moreover, current information on the 3D structure of proteins and their functions provide a possibility to understand the relevant molecular interactions between a ligand and a target macromolecule. As a consequence, a comprehensive study of drug structure–activity relationships can help identify a 3D pharmacophore model as an aid for rational drug design, as a pharmacophore model can be defined as 'an ensemble of steric and electronic features that is necessary to ensure the optimal supramolecular interactions with a specific biological target and to trigger (or block) its biological response', and a pharmacophore model can be established either in a ligand-based manner, by superposing a set of active molecules and extracting common chemical features that are essential for their bioactivity, or in a structure-based manner, by probing possible interaction points between the macromolecular target and ligands.

Molecular recognition (MR) is a general term designating non-covalent interactions between two or more compounds belonging to host-guest, enzyme-inhibitor and/or drug-receptor complexes. A rigorous approach to an MR study should involve the adoption of a computational method independent from the chemical intuition of the researcher. Drug design purposes prompt another challenging feature of such an ideal computational method, the ability to make sufficiently accurate thermodynamic predictions about the recognition process.

2. Molecular modelling methods and their usefulness

Molecular recognition is a central phenomenon in biology, for example, with enzymes and their substrates, receptors and their signal inducing ligands, antibodies and antigens, among others. Given two molecules with 3D conformations in atomic detail, it is important to know if the molecules bind to each other and, if it is so, what does the formed complex look like ("docking") and how strong is the binding affinity (that can be related to the "scoring"functions).

Molecules are not rigid. The motional energy at room temperature is large enough to let all atoms in a molecule move permanently. That means that the absolute positions of atoms in a molecule, and of a molecule as a whole, are by no means fixed, and that the relative location of substituents on a single bond may vary with time. Therefore, any compound containing one or several single bonds exists at every moment in many different conformers, but generally only low energy conformers are found to a large extent [Kund, 1997, as cited in Tóth et al., 2005].

The biological activity of a drug molecule is supposed to depend on one single unique conformation amongst all the low energy conformations, the search for this so-called bioactive conformation for compound sets being one of the major tasks in Medicinal Chemistry. Searching for all low energy conformations is possible with molecular modelling studies, since molecular modelling is concerned with the description of the atomic and molecular interactions that govern microscopic and macroscopic behaviours of physical systems. These molecular interactions are classified as: (a) bonded (stretching, bending and torsion), (b) non-bonded (electrostatic (including interactions with metals), van der Waals and π-stacking), and (c) derived, as they result from the previous ones (hydrogen bonds and hydrophobic effect).

Protein-ligand or, in general, molecule-molecule binding free energy differences can *a priori* be computed from first principles using free energy perturbation techniques and a full atomic detailed model with explicit solvent molecules using molecular dynamics simulations. However, these are computationally demanding. More affordable approaches use end-point molecular dynamic simulations and compute free energies accounting for solvent effects with continuum methods, such as MM-PBSA (molecular mechanics Poisson-Boltzman surface area) or MM-GBSA (generalized Born surface area) [Kollman et al., 2000; Wang et al., 2005]. One of the first approaches was comparative molecular field analysis (CoMFA) [Cramer et al., 1988], which enabled interpretation and understanding of enzyme active sites when the crystal structure was absent. However, this type of analysis was not possible until *in vitro* drug-drug interaction studies were widely used (through the 1990s).

2.1. Molecular mechanics, molecular dynamics and docking

Molecular mechanics (MM) is often the only feasible means with which to model very large and non-symmetrical chemical systems such as proteins and polymers. Molecular mechanics is a purely empirical method that neglects explicit treatment of electrons, relying instead on the laws of classical physics to predict the chemical properties of molecules . As a result, MM calculations cannot deal with problems such as bond breakage or formation, where electronic or quantum effects dominate. Furthermore, MM models are wholly system-dependent. MM energy predictions tend to be meaningless as absolute quantities, as the zero or reference value depends on the number and types of atoms and their connectivity, and so they are generally useful only for comparative studies. A force field is an empirical approximation for expressing structure-energy relationships in molecules and is usually a compromise between speed and accuracy.

Molecular mechanics have been shown to produce more realistic geometry values for the majority of organic molecules, owing to the fact that they are highly parameterised. Parameterisation of structures should be performed with care and non-"standard" molecules will need to have new parameters. This is usually done by analogy for bonded terms and assigning charges by a procedure consistent with the used force field.

There are many levels of theory in which computational models of 3D structures can be constructed. The overall aim of modelling methods is often to try to relate biological activity to structure. An important step towards this goal is to be able to compute the potential energy of the molecule as a function of the position of the constituent atoms. Once a method for evaluating the molecular potential energy is available, it is natural to search for an optimum molecular geometry by minimising the energy of the system. In a biological macromolecule, the potential energy surface is a complicated one, in which there are many local energy minima as well as a single overall energy minimum. All the energy minimisation algorithms commonly used have a marked tendency to locate only a local energy minimum that is close to the starting conformation. For a biological macromolecule, the number of conformations that have to be searched rises exponentially with the size of the molecule; hence, systematic searching is not a practical method for large molecules.

Molecular dynamics (MD) is a conformation space search procedure in which the atoms of a biological macromolecule are given an initial velocity and are then allowed to evolve in time according to the laws of Newtonian mechanics [van Gunsteren & Berendsen, 1977]. Depending on the simulated temperature of the system, the macromolecule can then overcome barriers at the potential energy surface in a way that is not possible with a minimisation procedure. One useful combination of molecular dynamics and minimisation schemes is a method known as simulated annealing [Kirkpatrick et al, 1983, Černý, 1985]. This method uses a molecular dynamics calculation in which the system temperature is raised to a high value to allow for a widespread exploration of the available conformational space. The system temperature is then gradually decreased as further dynamics are performed. Finally, a minimisation phase may be used to select a minimum energy molecular conformation.

One of the most important applications of molecular modelling techniques in structural biology is the simulation of the docking of a ligand molecule onto a receptor. These methods often search to identify the location of the ligand binding site and the geometry of the ligand in the active site, to get the correct ranking when considering a series of related ligands in terms of their affinity, or to evaluate the absolute binding free energy as accurately as possible. To select a force field and the adequate modelling methodology for a given task, it is important to appreciate the range of molecular systems to which it is applicable and the types of simulations that can be performed.

2.2. Most used existing force fields

AMBER (Assisted Model Building with Energy Refinement) developed by Kollman et al. [http://ambermd.org/] was originally parameterised specifically for proteins and nucleic acids [Weiner et al., 1984, 1986; Cornell et al., 1995], using 5 bonding and non-bonding terms along with a sophisticated electrostatic treatment and with no cross terms included. The results obtained with this method can be very good for proteins and nucleic acids, but less so for other systems, although parameters that enable the simulation of other systems have been published [for examples see Doshi & Hamelberg, 2009; Zgarbová et al., 2011].

CHARMM (Chemistry at HARward Macromolecular Mechanics) developed by Karplus et al. [http://www.charmm.org] was originally devised for proteins and nucleic acids [Brooks et al., 1983], and is now used for a range of macromolecules, molecular dynamics, solvation, crystal packing, vibrational analysis and QM/MM (quantum mechanics/molecular mechanics) studies. It uses five valence terms, one of which is electrostatic and is a basis for other force fields (e.g., MOIL [Elber et al., 1995]).

GROMOS (Groningen Molecular Simulation) developed at the University of Groningen and the ETH (Eidgenössische Technische Hochschule) of Zurich [http://www.igc.ethz. ch/GROMOS/index] is quite popular for predicting the dynamical motion of molecules and bulk liquids, also being used for modelling biomolecules. It uses five valence terms, one of which is electrostatic [van Gunsteren and Berendsen., 1977]. Its parameters are currently being updated [Horta et al., 2011].

MM1-4 (Molecular Mechanics) developed by Allinger [1976] are general purpose force fields for monofunctional organic molecules. The first version of this method was the MM1 [Allinger, 1976]. MM2 was parameterised for a lot of functional groups while MM3 [Allinger & Durkin, 2000; Allinger & Yan, 1993] is probably one of the most accurate ways of modelling hydrocarbons. MM4 is the latest version with several improvements [Allinger et al., 1996].

MMFF (Merck Molecular Force Field) developed by Halgren [1996] is also a general purpose force field mainly for organic molecules. MMFF94 [Halgren, 1996] was originally designed for molecular dynamics simulations, but has also been widely used for geometrical optimisation. It uses five valence terms, one of which is electrostatic and another is a cross term. MMFF was parameterised based on high level *ab initio* calculations. MMFF94 contains parameters for a wide variety of functional groups that arise in Organic and Medicinal Chemistry.

OPLS (Optimized Potential for Liquid Simulations) developed by Jorgensen at Yale [http://zarbi.chem.yale.edu] was designed for modelling bulk liquids [Jorgensen & Tirado-Rives, 1996] and has been extensively used for modelling the molecular dynamics of biomolecules. It uses five valence terms, one of which is an electrostatic term and none of them is a cross term.

TRIPOS (Sybil force field) is a commercial method designed for modelling organics and biomolecules. It is often used for CoMFA analysis and uses five valence terms, one of which is an electrostatic term.

CVFF (Consistent Valence Force Field) developed by Dauber-Osguthorpe is a method parameterised for small organic (amides and carboxylic acids, among others) crystals and gas phase structures [Dauber-Osguthorpe et al., 2004]. It handles peptides, proteins and a wide range of organic systems. It was primarily intended for studies of structures and binding energies, although it predicts vibrational frequencies and conformational energies reasonably well.

5. Selective formation of complexes

There are several examples of molecular modelling studies on complexes between cyclic receptors and ammonium ions, calixarenes [Choe & Chang, 2002] and crown ethers being the most used. As an example, it is noteworthy to mention the theoretical studies on calix[4]crown-5 and a series of alkyl ammonium ions [Park et al., 2007], having shown that the energy of complex formation depends on the number of amine groups in the alkyl chain as well as on the number of methylene groups between the primary and secondary amine groups, results that agree with experimental measurements. Although the calculations are performed under quite different conditions of vacuum compared with the experimental conditions of the phase system of chloroform-water, the binding properties of calixarene-type compounds towards alkyl ammonium ions have been successfully simulated, providing general and useful explanations for the molecular recognition behaviour.

Complex formation of compounds containing benzene rings with ammonium cations has also been theoretically studied using many computational techniques, including *ab initio* calculations [Kim et al., 2000]. It has been shown that two types of NH-aromatic π and CH-aromatic π interactions, which are important in biological systems, are responsible for binding, and that charged hydrogen bonds versus cation-π interaction is the origin of the high affinity and selectivity of novel receptors for NH_4^+ over K^+ ions [Oh et al., 2000]. Organic molecules complexed with metal cations have also been studied by MM2 molecular modelling [Mishra, 2010]. The search for metal ion selectivity is of interest in the field of biomimetic models of metalloenzymes and molecular modelling helps in the design of new ligands with this purpose [Kaye, 2011].

Molecular modelling has been used to suggest possible contributions of carrier effectivity and selectivity to complex formation in accordance with experimental results [Chipot et al., 1996; Ilioudis et al., 2005]. Our research group has evaluated the possible cation-receptor interactions involved in the complexes with ammonium and metal cations of selective carriers using the Amber force field with appropriate parameters developed by us. The complexation energies obtained are in reasonable agreement with experimental values, taking into account that complexation/decomplexation processes have a great influence on transport rates and are not equally favoured in cyclic and acyclic carriers [Campayo et al., 2004].

Both binding and selectivity in binding can be understood through the combined efforts of several non-covalent interactions, such as hydrogen bonding, electrostatic interactions, hydrophobic interactions, cation-π interactions, π-π stacking interactions and steric complementarity [Späth & König, 2010]. Formation of complexes is also possible in the case of neutral ligands. For example, the interactions between cholesterol and cyclodextrins have been theoretically studied to investigate their 1:1 and 1:2 complexes [Castagne et al., 2010], while the formation of stable complexes between trehalose and benzene compounds have been investigated by the general Amber force field (GAFF) and Gaussian 03 for MP2/6G-31G** calculation of atomic charges [Sakakura et al., 2011].

Docking of a ligand into a receptor may occur via an automated procedure [Subramanian et al., 2000] or manually [Filizola et al., 1999]. In both cases, docking is a combination of two components: a search strategy and a scoring function [Taylor et al., 2002]. The computational method MOLINE (Molecular Interaction Evaluation) was created to study complexes in an unbiased fashion [Alcaro et al., 2000]. It is based on a systematic, automatic and quasi-flexible docking approach that prevents the influence of the chemist's intuition on generating the configuration. This method has been used with acceptable results in studying inclusion complexes [Alcaro et al., 2004].

It would be adequate at this point to remember that testing the 'drug-receptor complexation' for a receptor model against available experimental data usually involves the use of site-directed mutagenesis experiments. This fact provides information on the amino acids involved in ligand binding and receptor activation. However, it should be noted that the results of mutagenesis studies are not necessarily related to receptor-ligand interactions. In fact, mutations can also alter the 3D structure of a receptor and therefore, modify the binding profile of a ligand by this mechanism. Besides that, efficient binding to a receptor does not guarantee that a ligand will produce a pharmacological action, given that the ligand may act as an agonist or antagonist.

6. Interaction of molecules with DNA

Anthracycline antibiotics such as doxorubicin and its analogues have been in common use as anticancer drugs for around half a century. There has been intense interest in the DNA-binding sequence specificity of these compounds in recent years, with the hope of identifying a compound that can modulate gene expression or exhibit reduced toxicity. Cashman and Kellog have studied models of binding for doxorubicin and derivatives [Cashman & Kellog, 2004], looking for sequence specificity and the effects of adding aromatic or aliphatic ring substituents or additional amino or hydroxyl groups. They performed a hydropathic interaction analysis using the HINT program (a Sybyl program module, Tripos Inc.) and four double base pair combinations. Interaction of some intercalators with two double DNA base pairs have also been studied with the density functional based tight binding (DFTB) method [Riahi et al., 2010], despite DFT methods being known to be inherently deficient in calculating stacking interactions, and the Amber force field and then AM1 to dock the intercalator between DNA base pairs [Miri et al., 2004].

Studies on sequence-selectivity of DNA minor groove binding ligands have shown that the most reliable results for AT-rich DNA sequences are obtained when MD simulations are performed in explicit solvent, when the data are processed using the MM-PB/SA approach, and when normal mode analysis is used to estimate configurational entropy changes [Shaikh et al., 2004; Wang & Laughton, 2009]. Use of the GB/SE model with a suitable choice of parameters adequately reproduces the structural and dynamic characteristics in explicitly solvated simulations in approximately a quarter of the computational time, although limitations become apparent when the thermodynamic properties are evaluated [Sands & Laughton, 2004]. Water molecules taking part in the complexation have been studied using

Doshi, U. & Hamelberg, D. (2009). Reoptimization of the AMBER force field parameters for peptide bond (omega) torsions using accelerated molecular dynamics. *J. Phys. Chem. B*, Vol. 113, No. 52, (December 2009), pp. 16590-16595. ISSN: 1089-5647

Durand, S., Dognon, J.P., Guiband, P., Rabbe, C. & Wipff, G. (2000). Lanthanide and alkaline-earth complexes of EDTA in water: a molecular dynamics study of structures and binding selectivities. *J. Chem. Soc., Perkin Trans. 2*, Vol. 2000, No. 4, (April 2000), pp. 705-714. ISSN: 1364-5471

Dzubiella, J., Swanson, J.M.J. & McCammon, J.A. (2006a). Coupling nonpolar and polar solvation free energies in implicit solvent models. *J. Chem. Phys.*, Vol. 124, (February 2006), pp. 084905. ISSN: 0021-9606

Dzubiella, J., Swanson, J.M.J. & McCammon, J.A. (2006b). Coupling hydrophobicity dispersion, and electrostatics in continuum solvent models. *Phys. Rev. Lett.*, Vol. 96, No. 8, (March 2006), pp. 087802. ISSN: 1079-7114

Elber, R., Rotberg, A., Simmerling, C., Goldstein, R., Li, H., Verkhivker, G., Keasar, C., Zhang, J. & Ulitsky, A. (1995). MOIL: A program for simulations of macromolecules. *Comp. Phys. Comm.*, Vol. 91, No. 1-3, (September 1995), pp. 159-189. ISSN: 1815-2406

Elguero, J. Alkorta, I., Claramunt, R.M., López, C., Sanz, D. & Santa María, D. (2009). Theoretical calculations of a model of NOS indazole inhibitors: Interaction of aromatic compounds with Zn-porphyrins. *Bioorg. Med. Chem.*, Vol. 17, No. 23, (December 2009), pp. 8027-8031. ISSN: 0968-0896

Fantoni, A.C. (2003). Molecular dynamics study of geometrical isomers of a pyridinocalix[4]arene in methanol solution: solvation and alkali metal cation binding properties. *J. Mol. Struct. (Theochem)*, Vol. 693, No. 1, (August 2003), pp. 1-6. ISSN: 0166-1280

Filizola, M., Carteri-Farina, M. & Perez, J.J. (1999). Molecular modeling study of the differential ligand-receptor interaction at the μ, δ and κ opioid receptors. *J. Comput. Aid. Mol. Des.*, Vol. 13, No. 4, (July 1999), pp. 397-407. ISSN: 1573-4951

Friesner, R.A., Banks, J.L., Murphy, R.B., Halgren, T.A., Klicic, J.J., Mainz, D.T., Repasky, M.P., Knoll, E.H., Shelley, M., Perry, J.K., Shaw, D.E., Francis, P. & Shenkin, P.S. (2004). Glide: A new approach for rapid, accurate docking and scoring. 1. Method and assessment of docking accuracy. *J. Med. Chem.*, Vol. 47, No. 7, (March 2004), pp. 1739-1749. ISSN: 0022-2623

Friesner, R.A., Murphy, R.B., Repasky, M.P., Frye, L.L., Greenwood, J.R., Halgren, T.A., Sanschagrin, P.C. & Mainz, D.T. (2006). Extra precision Glide: Docking and scoring incorporating a model of hydrophobic enclosure for protein-ligand complexes. *J. Med. Chem.*, Vol. 49, No. 21, (October 2006), pp. 6177-6196. ISSN: 0022-2623

Frølund, B., Jensen, L.S., Guandalini, L., canillo, C., Vestergarard, H.T., Kristiansen, U., Nielsen, B., Stensbøl, T.B., Madsen, C., Krogsgaard-Larsen, P. & Liljefors, T. (2005). Potent 4-aryl- or 4-arylalkyl-substituted 3-isoxazolol $GABA_A$ antagonists: synthesis, pharmacology, and molecular modeling. *J. Med. Chem.*, Vol. 48, No. 2, (January 2005), pp. 427-439. ISSN: 0022-2623

Galisteo, J., Navarro, P., campayo, L., Yunta, M.J.R., Gómez-Contreras, F., Villa-Pulgarin, J.A., Sierra, B.G., Mollinedo, F., Gonzalez, J. & García-España, E. (2010). Synthesis and cytotoxic activity of a new potential bisintercalator: 1,4-bis{3-[N-(4-chlorobenzo[g]phthalazin-1-yl)aminopropil]}piperazine. *Bioorg. Med. Chem.*, Vol. 18, No. 14 (July 2010), pp. 5301-5309. ISSN: 0968-0896

Gallicchio, E. & Levy, R.M. (2004). AGBNP: An analytical implicit solvent model suitable for molecular dynamics simulations and high-resolution modeling. *J. Comput. Chem.*, Vol. 25, No. 4, (March 2004), pp. 479-499. ISSN: 1096-987X

Gallicchio, E., Zhang, L.Y. & Levy, R.M. (2002). The SGB/NP hydration free energy model based on the surface generalized born solvent reaction field and novel nonpolar hydration free energy estimators. *J. Comput. Chem.*, Vol. 23, No. 5, (April 2002), pp. 517-529. ISSN: 1096-987X

Gevorkyan, A.A., Arakelyan, A.S., Esayan, V.A., Petrosyan, K.A. & Torosyan, G.O. (2001). Ionic character of the ammonium-counterion bond and catalytic activity of ammonium salts in elimination reactions. *Gen. Chem.*, Vol. 71, No. 8, (August 2001), pp. 1327-1328. ISSN: 1070-3632

Gilson, M.K. & Zhou, H.X. (2007). Calculation of protein-ligand binding affinities. *Annu. Rev. Biophys. Biomol. Struct.*, Vol. 36, (June 2007), pp. 21-42. ISSN: 1056-8700

Goodford, P.J. (1985). A computational procedure for determining energetically favorable binding sites on biologically important macromolecules. *J. Med. Chem.*, Vol. 18, No. 8, (August 1985), pp. 849-857. ISSN: 0022-2623

Halgren, T.A. (1996). Merck molecular force field. 1. Basis, form, scope, parameterización, and performanceof MMFF94. *J. Comput. Chem.*, Vol. 17, No. 5-6, (April 1996), pp. 490-519. ISSN: 1096-987X

Halgren, T.A., Murphy, R.B., Friesner, R.A., Beard, H.S., Frie, L.L., Pollard, W.T. & Banks, J.L. (2004). Glide: A new approach for rapid, accurate docking and scoring. 2. Enrichment factors in database screening. *J. Med. Chem.*, Vol. 47, No. 7, (March 2004), pp. 1750-1759. ISSN: 0022-2623

Hashimoto, S. & Ikuta, S. (1999). A theoretical study on the conformations, energetics, and solvation effects on the cation-π interaction between monovalent ions Li$^+$, Na$^+$ and K$^+$ and naphthalene molecules. *J. Mol. Struct. (Theochem)*, Vol. 468, No. 1-2, (August 1999), pp. 85-94. ISSN: 0166-1280

Hill, S.E. & Feller, D. (2000). Theoretical study of cation/ether complexes: 15-crown-5 and its alkali metal complexes. *Int. J. of Mass Spect.*, Vol. 201, No. 1-3, (December 2000), pp. 41-58, ISSN: 1387-3806

Hoops, S.C., Anderson, K.W. & Merz, K.M. Jr. (1991). Force field design for metalloproteins. *J. Am. Chem. Soc.*, Vol. 113, No. 22, (October 1991), pp. 8262-8270. ISSN: 0002-7863

Horta, B.A.C., Fuchs, P.F.J., van Gunsteren, W.F. & Hunenberger, P.H. (2011). New interaction parameters for oxygen compounds in the GROMOS force field: improved pure-liquid and solvation properties for alcohols, ethers, aldehydes, ketones, carboxylic acids and esters. *J. Chem. Theory Comp.*, Vol. 7, No. 4, (April 2011), pp. 1016-1031. ISSN: 1549-9626

Ilioudis, C.A., Bearpark, M.J. & Stead, J.W. (2005). Hydrogen bonds between ammonium ions and aromatic rings exist and have key consequences on solid-state and solution phase properties. *New J. Chem.*, Vol. 29, No. 1, (January 2005), pp. 64-67. ISSN: 1144-0546

Roumen, L., Peeters, J.W., Emmen, J.M.A., Bengels, I.P.E., Custers, E.M.G., de Gooyer, M, Plate, R., Pieterse, K., Hilbers, P.A.J., Smits, J.F.M., Vekemans, J.A.J., Leysen, D., Ottenheijm, H.C.J., Janssen, H.M. & Hermans, J.J.R. (2010). Synthesis, biological evaluation, and molecular modeling of 1-benzyl-1H-imidazoles as selective inhibitors of aldosterone synthase (CYP11B2). *J. Med. Chem.*, Vol. 53, No. 4, (February 2010), pp. 1712-1725. ISSN: 0022-2623

Sakakura, K., Okabe, A., Oku, K. & Sakurai, M. (2011). Experimental and theoretical study on the intermolecular complex formation between trehalose and benzene compounds in aqueous solution. *J. Phys. Chem. B*, Vol. 115, No. 32, (August 2011), pp.9823-9830. ISSN: 1089-5647

Sanchez-Moreno, M., Sanz, A.M., Gómez-Contreras, F., Navarro, P., Marín, C., Ramírez-Macías, I., Rosales, M.J., Olmo, F., García-Aranda, I., Campayo, L., Cano, C., Arrebola, F. & Yunta, M.J.R. (2011). *In vivo* Trypanosomicidal activity of imidazole-or pyrazole-based venzo[g]phthalazine derivatives against acute and chronic phases of chagas disease. *J. Med. Chem.*, Vol. 54, No. 4, (February 2011), pp. 970-979. ISSN: 0223-5234

Sands, Z.A. & Laughton, C.A. (2004). Molecular dynamics simulations of DNA using the generalized Born solvation model: quantitative comparisons with explicit solvation results. *J. Phys. Chem. B*, Vol. 108, No. 28, (July 2004), pp. 10113-10119. ISSN: 1089-5647

Shaikh, S.A., Ahmed, S.R. & Jayaram, B. (2004). A molecular thermodynamic view of DNA-drug interactions: A case study of 25 minor-groove binders. *Arch. Biochem. Biophys.*, Vol. 429, No. 1, (September 2004), pp. 81-99. ISSN: 0003-9861

Shivakumar, D., Williams, J. Wu, Y., Damm, W., Shelley, J. & Sherman, W. (2010). Prediction of absolute solvation free energies using molecular dynamics free energy perturbation and the OPLS force field. *J. Chem. Theory Comput.*, Vol. 6, No. 5, (May 2010), pp. 1509-1519. ISSN: 1549-9626

Shoichet, B.K. & Kuntz, I.D. (1996). Predicting the structure of protein complexes: a step in the right direction. *Chem. and Biol.*, Vol. 3, No. 3, (March 1996), pp.151-156. ISSN: 1074-5521

Silva, S.J. & Jayasundera, K. (2002). Quantitative structure activity relationships for guanidiniothiazole carboxamides using theoretically calculated molecular descriptors. *J. Natn. Sci. Found. Sri Lanka*, Vol. 30, No. 3-4, (December 2002), pp. 171-184. ISSN: 1391-4588

Simonson, T. (2001). Macromolecular electrostatics: continuum models and their growing pains. *Curr. Opin. Struct. Biol.*, Vol. 11, No. 2 (April 2001), pp. 243-252. ISSN: 0959-440X

Slickers, P., Hillebrand, M., Kittler, L., Löber, G. & Sühnel, J. (1998). Molecular modeling and footprinting studies of DNA minor groove binders: bisquaternary ammonium heterocyclic compounds. *Anti-Cancer Drug Des.*, Vol. 13, No. 5, (September 1998), pp. 463-488. ISSN: 0266-9536

Soteras, I., Morreale, A., López, J.M., Orozco, M. & Luque, F.J. (2004). Group contributions to the solvation free energy from MST continuum calculations. *Braz. J. Phys.*, Vol. 34, No. 1, (March 2004), pp. 48-57. ISSN: 1678-4448.

Späth, A. & König, B. (2010). Molecular recognition of organic ammonium ions in solution using synthetic receptors. *Beilstein J. Org. Chem.*, Vol. 6, No. 32, (April 2010), pp. 1-111. ISSN: 1860-5397

Srinivas, E., Murthy, J.N., Rao, A.R.R. & Sastry, G.N. (2006). Recent advances in molecular modeling and medicinal chemistry aspects of phosphor-glycoprotein. *Curr. Drug Metabol.*, Vol. 7, No. 2, (February 2006), pp. 205-217. ISSN: 1389-2002

Stoika, I., Sadiq, S.K. & Coveney, P.V. (2008). Rapid and accurate prediction of binding free energies for saquinavir-bound HIV-1 proteases. *J. Am. Chem. Soc.*, Vol. 130, No. 8, (February 2008), pp. 2639-2648. ISSN: 0002-7863

Subramanian, G., Paterlini, M.G., Portoghese, P.S. & Ferguson, D.M. (2000). Molecular docking reveals a novel binding site model for fentanyl at the μ-opioid receptor. *J. Med. Chem.*, Vol. 43, No. 3, (February 2000), pp. 381-391. ISSN: 0022-2623

Taylor, R.D., Jewsbury, P.J. & Essex, J.W. (2002). A review of protein-small molecule docking methods. . *J. Comput. Aid. Mol. Des.*, Vol. 16, No. 3, (March 2002), pp. 151-166. ISSN: 1573-4951

Tóth, J., Remko, M. & Nagy, M. (2005). The ability of molecular modeling methods to reproduce the structure of flavonoids. *Acta Facul. Pharm. Univ. Comenianae*, Vol. LII, (2005), pp. 218-225. ISSN: 0301-2298

Tschammer, N., Elsner, J. Goetz, A., Ehrlich, K., Schuster, S., Ruberg, M., Kühhorn, J., Thompson, D., Whistler, J., Hübner, H. & Gmeiner, P. (2011). Highly potent 5-aminotetrahydropyrazolopyridines: enantioselective dopamine D_3 receptor binding, functional selectivity, and analysis of receptor-ligand interactions. *J. Med. Chem.*, Vol. 54, No. 7, (April 2011), pp. 2477-2491. ISSN: 0022-2623

van Gunsteren, W.F. & Berendsen, H.J.C. (1977). Algorithms for macromolecular dynamics and constraint dynamics. *Mol. Phys.*, Vol. 34, No. 5, (August 2006), pp.1311-1327. ISSN: 1362-3028

Varnek, A. Wipff, G., Bilyk, A. & Harrowfield, J.M. (1999). Molecular dynamics and free energy perturbation studies of Ca^{2+}/Sr^{2+} complexation selectivities of the macrocyclic ionophores DOTA and TETA in water. *J. Chem. Soc. Dalton Trans.*, Vol. 1999, No. 23, (December 1999), pp. 4155-4164. ISSN: 1472-7773

Veal, J.M., Li, X., Zimmerman, S.C., Lambenmon, C.R., Cory, M., Zon, G. & Wilson, W.D. (1990). Interaction of a macrocyclic bisacridine with DNA. *Biochem.*, Vol. 29, No. 49, (December 1990), pp. 10918-10927. ISSN: 0006-2960

Venskutonyte, R., Butini, S., Coccone, S.S., Gemma, S., Brindisi, M., Kumor, V., Guarino, E., Maramai, S., Amir, A., Valades, E.A., Frydenvang, K., Kastrup, J.S., Novellino, E., Campiani, G. & Pickering, D.S. (2011). Selective kainite receptor (Gluk1) ligands structurally based upon 1*H*-cyclopentapyrimidin-2,4(1*H*,3*H*)-dione: Synthesis, molecular modeling, and pharmacological and biostructural characterization. *J. Med. Chem.*, Vol. 54, No. 13, (July 2011), pp. 4793-4805. ISSN: 0022-2623

Viswanadhan, V.N., Ghose, A.K., Sing, U.C. & Wendoloski, J.J. (1999). Prediction of solvation free energies of small organic molecules:additive-constitutive models based on molecular fingerprints and atomic constants. *J. Chem. Inf. Comput. Sci.*, Vol. 39, No. 2, (March 1999), pp. 405-412. ISSN: 0095-2338

Wang, J., Kang, X., Kuntz, I.D. & Kollman, P.A. (2005). Hierarchical database screenings for HIV-1 reverse transcriptase using a pharmacophore model, rigid docking and MM-PB/SA. *J. Med. Chem.*, Vol. 48, No. 8, (April 2005), pp. 2432-2444. ISSN: 0022-2623

Wang, H. & Laughton, C.A. (2009). Evaluation of molecular modeling methods to predict the secuence-selectivity of DNA minor groove binding ligands. *Phys. Chem. Chem. Phys.*. Vol. 11, No. 45, (December 2009), pp. 10722-10728. ISSN: 1463-9076

Wang, H. & Laughton, C.A. (2010). Molecular modeling mrthods to quantitative drud-DNA interactions, In: *Drug-DNA interaction protocols, Methods in molecular biology*, Vol. 613, pp. 19-31, Humana Press, Germany. ISBN: 978-1-60327-417-3

Weiner, S.J., Kollman, P.A., Case, D.A., Singh, V.C., Ghio, C., Alagona, G., Profeta, S. Jr. & Weiner, P. (1984). A new force field for molecular mechanical simulation of nucleic acids and proteins. *J. Am. Chem. Soc.*, Vol. 106, No. 3, (February 1984), pp. 765-784. ISSN: 0002-7863

Weiner, S.J., Kollman, P.A., Nguyen, D.T. & Case, D.A. (1986). An all atom force field for simulations of proteins and nucleic acids. *J. Comput. Chem.*, Vol. 7, No. 2, (April 1986), pp. 230-252. ISSN: 1096-987X

Wilson, C., Mace, J.E. & Agard, D.A. (1991). A computational method for designing enzymes with altered substrate specifity. *J. Mol. Biol.*, Vol. 220, No. 2, (July 1991), pp. 495-506. ISSN: 0022-2836

Woods, R.J., Dwek, R.A., Edge, C.J. & Fraser-Reid, B. (1995). Molecular mechanical and molecular dynamical simulations of glycoproteins and oligosaccharides. 1. GLYCAM_93 parameter development. *J. Phys. Chem.*, Vol. 99, No. 11, (March 1995), pp. 3832-3846. ISSN: 0022-3654

Yang, L., Tan, C., Hsieh, M-J., Wang, J., Duan, Y., Cieplak, P., Caldwell, J., Kollman, P.A. & Luo, R. (2006). New generation Amber united-atom force field. *J. Phys. Chem. B*, Vol. 110, No. 26, (July 2006), pp. 13166-13176. ISSN: 1089-5647

Zacharias, N. & Dougherty, D.A. (2002). Cation-π interactions in ligand recognition and catalysis. *Trends Pharm. Sci.*, Vol. 23, No. 6, (June 2002), pp. 281-287. ISSN: 0165-6147

Zgarbová, M., Otyepka, M., Sponer, J., Mládek, A., Banés, P., Cheatham, T.E. III & Jurecka, P. (2011). Refinement of the Cornell et al. nucleic acids force field based on reference quantum chemical calculations of glycosidic torsion profiles. *J. Chem. Theory Comp.*, Vol. 7, No. 9, (September 2011), pp. 2886-2902, ISSN: 1549-9626

Structural Bioinformatics

On the Assessment of Structural Protein Models with ROSETTA-Design and HMMer: Value, Potential and Limitations

León P. Martínez-Castilla and Rogelio Rodríguez-Sotres

Additional information is available at the end of the chapter

http://dx.doi.org/10.5772/47842

1. Introduction

The prediction of the three-dimensional structure of a protein, starting with the amino acid sequence, is still an unsolved issue. However a number of important advancements have been made and some methods offer solutions to this problem, specially when the target sequence has homologues whose structure has been determined. In any case, it is important to evaluate the quality of the prediction, as none of the methods offers assurance of success. The ROSETTA-design-HHMer (Rd.HMM) protocol stands out among the current quality assessment methods, because it offers evidence of the biological appropriateness of the prediction. In addition, Rd.HMM can be used to guide the modeling process towards the improvement of the model's quality. This chapter deals with the principles behind this protocol and gives practical advice on how to use the Rd.HMM to evaluate the quality of a three-dimensional modeled structure of a protein, and how to use the information to improve the model. The limitations of the protocol are also discussed.

2. The folding problem is a NP-hard problem involving a degenerate informational code

As implied by the well-known Levinthal paradox (Levinthal, 1968), a full exploration of the entire conformational space theoretically available to a protein is out of the reach of current computational techniques. Equally unaccessible to nature is the sequence space available to polypeptide chains (Kono & Saven, 2001). Currently, the amount of available protein structures (the PDB) represents a fraction of the known protein amino acid sequences, and if the available sample is grouped in terms of different folds, the diversity in the PDB is even smaller. In addition, protein structure and function can tolerate a significant number of

mutations. Both facts suggest an important degree of degeneracy between the information in polypeptide sequences and the associated code leading to their native structure (Bowie et al., 1990). In other words, the so-called folding code is degenerate.

However, even if the number of protein structural folds is smaller that the sequence space, the folding problem is still unsolved, because exploring the total number of conformations available to a protein or its energy landscape are NP-hard problems (Hart & Istrail 1997), and because the available methods to calculate the energy of a protein conformation imply a large systematic error (Faver et al., 2011).

The above facts set forth the intractability of solving the problem through an exhaustive search. Nevertheless, proteins in nature do reach a native structure in short times, and finding a native-like solution to the three-dimensional structure of a protein may not require a full examination of the conformational space, or its corresponding energy landscape. In fact, recent years have seen important progress in the search for solutions to the protein folding problem (Dill et al., 2008).

3. The problem of quality scoring for 3D models

In theory, the native three-dimensional structure of a protein must lie at an energy minimum, underneath all accessible intermediates with near-native fold. However, an accurate calculation of the energy for a protein conformation requires quantum chemical calculations. Properties such as electron-electron correlation, charge transfer, polarization, and bond break/formation, including proton exchange, involve quantum mechanical effects and cannot be correctly described using the equations of classical physics. The relevance of quantum mechanics for accurate energy calculations of protein-ligand complexes and protein conformations have been recently demonstrated (Raha and Merz, 2005). Numerical approximations to the electronic state of a multielectronic system have been developed for a variety of system up to date. But only a few simplified solutions, implying low-precision, can tackle an electronic macromolecular system, and even these demand a large amount of computational resources (He & Merz, 2010). The common simplifications, based on molecular mechanics, do carry a systematic error that precludes the accurate finding of the true native energy minimum (Faver et al., 2011).

Many methods have been proposed to model the three-dimensional structure of proteins starting from their amino acid sequence. Based on their use of experimental structural information, these methods can be classified into comparative modeling or *ab initio* methods.

Because rating the success of any method requires an impartial judge to be trustworthy, the scientific community implemented the contests for CRITICAL ASSESMENT OF THE STRUCTURE OF PROTEINS (CASP) (Kryshtafovych et al., 2009). In such contests, the judges are computer algorithms, which compare a 3D-structure solved by an experimental method (but yet unpublished) to a 3D-model predicted by a CASP contestant. The comparative modeling strategies have had a remarkable degree of success in the prediction of 3D-structures of soluble proteins, with the amino acid sequence as starting information.

Comparative modeling exploits the wealth of experimental structural information nowadays available for proteins (Rose et al., 2011), and relies on powerful sequence alignment algorithms (Wallace et al., 2005). In CASP contests, comparative modeling servers, such as I-TASSER (Roy et al., 2010), ROBETTA (Kim et al., 2004) and SAM-T08 (Karplus, 2009), have achieved a high success rate in their predictions for protein 3D-structures of low to intermediate difficulty (as defined by the CASP staff). Yet, one mayor limitation in these methods lies in the strategies used to match each amino acid in a target sequence to its corresponding best hosting spot in the 3D-structure of the template and, again, this is a NP-complete problem (Lathrop, 1994).

In *ab initio* methods, the laws of physics and chemistry and/or artificial intelligence are used to generate a prediction for a native-like folding solution of a protein with known amino acid sequence (Dill et al., 2008). While *ab initio* methods have been less successful than comparative modeling, these are the only choice if no suitable homologous 3D-template is available, for a given amino acid sequence (Kryshtafovych et al., 2009).

The above considerations are all fine when the question is to grade the methods and chose the one with highest success rate, but to date, no single method gives the correct answer every time. Yet, the final aim of such methods is to produce good native-like protein 3D-predictions, when experimental X-ray or NMR data are not available. How then is it possible to set apart models with wrong fold assignment, from those with a correct fold assignment, but with a mistraced sequence to 3D-fold alignment (Luthy et al., 1992)? Is it possible to identify cases where the fold assignment and the alignment are adequate, but the solution to the atom repacking of replaced amino acids is deficient? These questions lie behind the quality assessment of a protein 3D-structure prediction.

The quality assessment is of particular relevance in cases where a suitable 3D-template cannot be found, because the predicted 3D-model cannot be compared back the starting template. Again, this problem can be tackled with a number of strategies, and most of them have been implemented as computer software programs, and their validity tested at the CASP contests (Shi et al., 2009).

Quality assessment methods for the predicted 3D-structures of proteins can be classified according to their underlying principles:

i.   Physics-based methods use the regularities in chemical structures and the laws of physics and chemical bonding to find how much a 3D-structure deviates from the known canonical values. These methods may come in the form of force-fields and they report energies (Hu & Jiang, 2010), or may seek for abnormalities in geometrical and chemical features such as bonding lengths, bonding angles, dihedral torsion values, charge-charge distances and so on (Rodriguez et al. 1998).

ii. Statistics-based methods use the known 3D-structures to generate a set of probability distributions for a number of features of the experimentally solved structures. These distributions can be used as reference to judge the quality of a prediction. When these probability distributions are transformed into energies, using the Boltzmann law, the

more detailed description of this kind of problems in protein modeling and how to fix them can be found elsewhere (Chavelas Adame et al., 2011; Rosales-León et al., 2012).

3. Build many replicates of your PDB file with a random assignment of amino acid sequence. This can be done in two ways:

a. *With VMD.* Use the *atomselect* command to select the backbone atoms of all residues, one at a time, and change the residue name to any amino acid selected at random. In the C-terminal residue make sure to include the OXT atom in your selection. Then select all backbone atoms including the OXT and save the file. A script to do this with VMD can be requested to the corresponding author.

b. *With Rd.* Prepare several Rosetta input *resfile* with a tag *PICKAA X* (replace X for a random 1-letter amino acid code) for every position in the PDB file of interest. Then run Rd, once for each *resfile* you made, with the following command:

rosetta.gcc -s 1QYS.pdb-design -fixbb -chain A -resfile 1QYSaa.res-ndruns 1 -pdbout 1QYSAaa

Here, we assume 1QYS.pdb to be the starting PDB file, *A* to be the subunit of interest, *1QYSaa.res* is the *resfile*, and your result will be named *1QYSAaa_0001.pdb* (depending on the version, you will also need a paths.txt in your folder, or the path to the rosetta database should be indicated in the command line). In Rd v. 3.1, the *resfile* format and some command line options have changed (check Rosetta documentation for details).

This step can be repeated at will, to create many sequence-randomized PDB files, but in our experience, at least 10 are needed for a reliable HMM.

4. Rebuild each sequence-randomized PDB file with Rd. First you will need a *resfile* with the tag *ALLAA* (1QYSall.res), and a text file (pdb4rbld.lst) containing the names of all the sequence-randomized PDB file created in the previous step, one per line. Then, rebuild each input file 29 times using the command:

rosetta.gcc -design -fixbb -chain A-l pdb4rbld.lst -resfile 1QYSall.res -ndruns 29 -pdbout

You can generate many rebuilt PDBs per input file, but you need at least 100 sequences in the end to produce a representative HMM. In our experience, a better exploration of the sequence space results from many sequence-randomized input files and between 10 to 30 rebuilt PDBs for each input PDB file.

5. Extract the amino acid sequence for each rebuilt PDB file and save it in a text file (*1QYSA_a-O.fas*), in fasta format. This file represents an alignment, though a sequence alignment software is not necessary, due to the reasons commented at the end of section 5. Then use HMMer to prepare a hidden Markov model of your sequence alignment:

hmmbuild --informat afa 1QYSA_a-O.hmm 1QYSA_a-O.fas

If you are using HMMer v. 2.0, you need to calibrate your model with:

hmmcalibrate 1QYSA_a-O.hmm

6. Search a sequence database (*i.e.* NCBI-nr) with:

hmmsearch -E 100 -Z 10000000 1QYSA_a-O.hmm path2db/nr

Here a local copy of the nr is assumed to be in your system in fasta format. The -Z flag will scale the E-values to 10 million sequences. This is recommended to make E-values comparable, because E-values are linearly dependent on the size of the sequence database searched. The default E value is 10, but here it was set to 100 to lower the search threshold.

```
A  # hmmsearch3 :: search profile(s) against a sequence database
   # HMMER 3.0 (March 2010); http://hmmer.org/
   # query HMM file:                1qys_cf.a-O.hmm
   # target sequence database:      /busr/dbr/nr
   # sequence reporting threshold:  E-value <= 100 ...

B  Scores for complete sequences (score includes all domains):
      --- full sequence ---   --- best 1 domain ---   -#dom-
      E-value  score  bias   E-value  score  bias   exp  N  Sequence                       Description
      -------  -----  -----  -------  -----  -----  ---- --  --------                       -----------
      6.3e-11   51.4   1.5   7.6e-11   51.2   1.0    1.2  1  gi|39654745|pdb|1QYS|A         Chain A...
      2.5e-07   39.8   0.0   2.7e-07   39.7   0.0    1.0  1  gi|118137815|pdb|2GJH|A        Chain A...
      4.5e-05   32.5   0.4   5.4e-05   32.3   0.2    1.1  1  gi|196049603|pdb|2JVF|A        Chain A...
      ------ inclusion threshold ------
         0.16   21.1   0.2   3.3e+02   10.3   0.0    2.3  2  gi|294496314|ref|YP_003542807.1|   hypothetical prot
      ...
C  Domain annotation for each sequence (and alignments):
   >> gi|39654745|pdb|1QYS|A  Chain A, Crystal Structure Of Top7: A Computationally Designed Protein With A Novel Fold
      #    score  bias  c-Evalue  i-Evalue hmmfrom  hmm to    alifrom  ali to    envfrom  env to     acc
      ---  ------ ----- --------- --------- ------- -------   ------- -------   ------- -------    ----
      1 !   51.2   1.0   1.7e-15   7.6e-11       4      91 .]       6      94 ..       3      94 .. 0.87

   Alignments for each domain:
     == domain 1    score: 51.2 bits;  conditional E-value: 1.7e-15
             1qys_cf.a-O2   4 iiviikDdndvlilvffvDGGierekvrek.ikiikylnalvveiaidseePskAiefakklyqfflelGFtDivivFdGtrvdvkGvL 91
                              + v i D+   +   ++v    e +kv+++    ik + a  v+i+i ++  ++A++fa +l + f elG+ Di ++FdG  v+v+G+L
     gi|39654745|pdb|1QYS|A   6 VQVNIDDNGKNFDYTYTVTTESELQKVLNElXDYIKKQGAKRVRISITARTKKEAEKFAAILIKVFAELGYNDINVTFDGDTVTVEGQL 94
                              556666666677777788888899999865157 9***********************************************************97 PP
   ...
   ...
   >> gi|196049603|pdb|2JVF|A  Chain A, Solution Structure Of M7, A Computationally-Designed Artificial Protein
      #    score  bias  c-Evalue  i-Evalue hmmfrom  hmm to    alifrom  ali to    envfrom  env to     acc
      ---  ------ ----- --------- --------- ------- -------   ------- -------   ------- -------    ----
      1 !   32.3   0.2   1.2e-09   5.4e-05       7      88 ..      12      93 ..       6      96 .] 0.84

   Alignments for each domain:
     == domain 1    score: 32.3 bits;  conditional E-value: 1.2e-09
             1qys_cf.a-O2   7 iikDdndvlilvffvDGGierek.vrekikiikylnalvveiaidseePskAiefakklyqfflelGFtDivivFdGtrvdvk 88
                              i +D +++ i + +   G e e+ ++e  k +    a  v+i+i +e+ ++A+e  + + +   +lG++Di +  +Gt v+++
     gi|196049603|pdb|2JVF|A 12 IQRDGQEIEIDIRVS-TGKELERaLQELEKALARAGARNVQITISAENDEQAKELLELIARLLQKLGYKDINVRVNGTEVKIE 93
                              556778888887655.566666615666789999**********************************************986 PP
```

Figure 4. An HMMer search output result. The search was done using an Rd.HMM from Top7 (PDB id. 1QYS) and the NCBI-nr as database. (A) heading, (B) scores, (C) domain-parsed scores and alignments. The statistics at the end and some information was removed for brevity. The format is as in HMMer 3.0.

An extract of the results from a typical search is presented in figure 4. The HMMer search output will report three sections: (a) Heading, (b) scores for complete sequences, (c) domain parsing, alignments and statistics. As it can be seen, according to the information in the Rd.HMM from Top7, the Top7 amino acid sequence fits into the Top7 3D-atomic coordinates (1QYS). The most relevant sections are the scores and the alignment sections. Notice how this X-ray solved 3D-stucture reports an HMM score of 51.2, matching the sequence from amino acid 3 to 94, that gives a ratio of 0.56, close to the 0.6 average value for X-ray solved structures. The reason behind the relationship is not simple, but it holds for most X-ray solved structures (with a few exceptions) (Martínez-Castilla & Rodríguez-Sotres 2010). The second hit is the C-terminal fragment of Top7 solved by NMR, the score is 39.8 for a fragment of length 50 (ratio = 0.79). This last score is higher than the score for the complete sequence, because as shown in figure 4, the C-terminus has higher proportion of local coincidences to the HMM. In the alignment to the full sequence, the contribution of the N-terminus lowers the overall score. The alignment for the C-terminal fragment was omitted because it is identical to the 1QYS alignment from position 44 to 91 (Fig. 4C). The alignment shows a consensus for the hidden Markov model, as a reference, then the sequence found aligned separated by an intermediate mask. Uppercase letters indicate strong conservation, lower case letter conservative changes and plus sign a positive local score. The lower line, absent in HMMer 2 is the encoded posterior probability (d=0...9,*; * equals 9.5), where the approximate value of posterior probability for each site is given by equation (1).

$$pp = d \times 0.1 + 0.025 \tag{1}$$

The final hit in the search in figure 4 is the M artificial protein. This protein was designed with Rosetta-design using the same Top7 folding. Its sequence is different, but it belongs to the same family of Rd proteins. The score is smaller than for Top7 (ratio of 32/(88-7), or 0.395), but still above 0.3 and with high statistical significance. Although Rd was used to design these proteins, the concordance reveals the robustness of the amino acid assignment made by Rd, and gives further support to the structurally aware nature of the Rd.HMM alignments.

The alignment is very useful to protein modeling, because it reveals the distribution of coincident regions between the 3D-atomic coordinates of the backbone and the amino acid sequence in the database. The following features are to be taken into account:

a. Frame shift. If the residue number in the 3D-structure has an offset relative to the amino acid numbering in the sequence, either from the beginning, or starting at some intermediate site; this is usually a sign of a wrong threading of the model and the template during the modeling step. In the example, there is a difference in amino acid numbering, but this is not a frame shift, as the first residue solved in the PDB is ASP-3, corresponding to node one in the HMM, then the first 3 HMM nodes did not match the Top7 sequence and were discarded by HMMer search making the first match to residue 6, at HMM node 4.

b. Insertion/deletions. An insertion in the sequence appears as a dot in the HMMer consensus, a deletion as dashes in the sequence found. Such changes are expected if the sequence is a homologue, and not the natural sequence that corresponds to the 3D-structures analyzed with Rd.HMM. They may occur also when the PDB file has some missing amino acids (this happens frequently, due to experimental limitations of X-ray crystalography). If so, you expect this insertions to match the missing amino acids. For *in silico* modeled structures this means a local threading error, or a local defect in the model.

c. Distribution of conserved sites. The higher the number of conserved sites, the better the model. However, some strained conformations have lower energy for glycine, proline and asparagine than for every other amino acid and these residues tend to appear as strongly conserved (Uppercase letters in the mask line, and in the Rd.HMM consensus). If the sequence conservation observed is dominated by these residues, you model may be wrong, even if your score has statistical significance.

9. Guiding the 3D-modeling of proteins with Rd-HMMer

There are many publications describing different approaches to the solution to the protein folding problem (Roy et al., 2010; Kim et al., 2004; Karplus, 2009; Melo & Feytmans, 1998) but most of them focus on the theory, or present a technical treatment. Fisher and Sali published a practical guide to the use of the popular modeling software MODELLER (Fiser & Sali 2003), where many useful hints are given. Recently, Chavelas-Adame and coworkers published a guide with emphasis on the use of open software [45]. The present account will not attempt to repeat the work, and only the most important conclusions are given here:

a. Many servers and software programs are available to aid the comparative modeling of proteins (Roy et al., 2010; Kim et al., 2004; Karplus, 2009; Melo & Feytmans, 1998; Rosales-León et al., 2012; Fiser & Sali 2003), some options are available for *ab initio* modeling (Kryshtafovych et al., 2009; Kim et al., 2004; Srinivasan et al., 2004; Xu & Zhang, 2012), and this list is far form complete. None of them achieves 100% success, and even the most successful can fail where other, usually less reliable, may succeed (Kryshtafovych et al., 2009; Melo & Feytmans, 1998).

b. A model is fundamentally wrong when the folding pattern in the model bears little or no relationship to the true native fold. Some models may offer a good approximation, but have wrong geometrical, sterical and/or chemical features at some locations, *i.e.* the bond lengths, angles, sidechain-sidechain contact distances and orientation may have important deviations from the expected values found in known chemical structures. This last kind are usually designated as unrefined models.

c. Unrefined models can be recognized with various energy scoring strategies (Luthy et al., 1992; Shi et al., 2009; Hu & Jiang, 2010; Melo & Feytmans, 1998); and can be corrected through the use of molecular mechanics software (Rosales-León et al., 2012; Fiser & Sali 2003), though this approach has limitations, as mentioned before (Faver et al., 2011; Hu & Jiang, 2010; Melo & Feytmans, 19985).

d. Wrong models instead may frequently be deceitful, because, due to their systematic error [5], a molecular mechanics force-field may report a low energy value, as long as the chemical and geometrical details are well refined. Rd.HMM offers a solution to this problem, because these models will produce an HMM search report with no hits, or will score sequences, other than the modeling target (Chavelas Adame et al., 2011; Rosales-León et al., 2012).

e. The analysis of the Rd.HMM search report may help in the identification of errors in the alignment between the target amino acid sequence and the template selected for comparative modeling. If you find a frame-shift or an unexpected insertion/deletion pair, you can use the HMM search alignment and realign the target sequence and the template. MODELLER is a very good choice for that aim (Fiser & Sali 2003). In addition, a wrongly threaded model can be recycled by replacing the consensus sequence with the PDB sequence in the model (which is the target sequence), and producing a target to target alignment, with the insertions and deletions suggested by HMMer. MODELLER can then be used to generate new models. This last procedure is only recommended if your HMMer score is positive and has good statistical significance, for otherwise, the structural inaccuracy of the Rd.HMMs becomes a serious issue.

f. Comparative modeling has been extended thanks to methods able to find templates with low sequence homology to the target (Wallace et al., 2005; Karplus, 2009). But sometimes the selected template is too distant. If the Rd.HMM of the candidate structures are obtained, these can be use to score the target sequence. The resulting scores, statistical significance and the alignment may guide your template selection. However, if the Rd.HMM of a template candidate gives a negative score, and still you decide to use it, do not trust the Rd.HMM alignment without further improvement using other tools, as it may be seriously flawed.

g.   Finally, if you use the ROSETTA suite or the ROBETTA server to produce your models, these structures are expected to have a ROSETTA-like bias, *i.e.* their Rd.HMM scores will increase and a good model with this bias is expected to have a ratio of HMMer score to sequence length close to one. While in models produced with other software a Rd.HMM score ratio of 0.3 is acceptable, in a ROSETTA produced model this score is low and may reflect important flaws. Look at the alignment carefully, as recommended in the previous section.

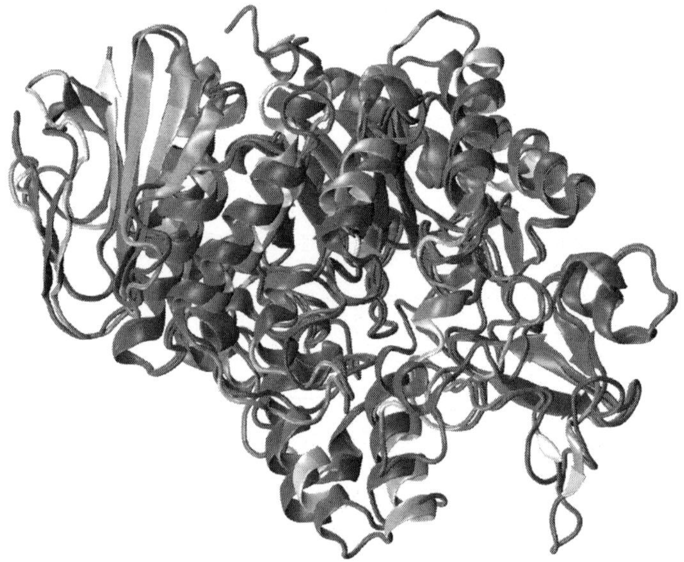

Figure 5. Comparison of the yeast α-glucosidase model produced included in the publication by Brindis et al (Brindis et al., 2011), with the X-ray solved structure of its homologue, the yeast isomaltase (Yamamoto et al., 2010). The isomaltase is shown as blue cartoons and the α-glucosidase cartoons are colored according to the amino acid rmsd from isomaltase, ranging from very low (blue) to intermediate (white) to high (red).

As an example of the advantages of Rd.HMM, we refer to two cases of recent success. Brindis and coworkers (Brindis et al., 2011) analyzed the effects of a natural product on α-glucosidase. This work reports a model for the budding yeast α-glucosidase used to analyze the molecular grounds for the (Z)-3-butylidenephthalide inhibitory action. In the preparation of the model, Rd.HMM allowed to detect a threading problem (insertion/deletion pair) in one β-strand in the core of the model. While the sheet was slid only a few Å from its position, the contact with neighboring strands completely distorted the chemical interaction network affecting the model stability. The correction of this problem and the use of molecular dynamics simulations led to a well refined and reliable model with a good Rd.HMM score. A few months later (when the paper was in press) the 3D-structure of a close homologue (isomaltase) was released (Yamamoto et al., 2010). The X-ray data corroborated the model quality, as the model core backbone has an rmsd of 1.81 Å form the experimental data. Figure 5 shows a superposition of both structures colored by backbone rmsd form blue (low) to white (medium) to red (high).

In a second example, the 3D-structure of two isoforms of plant inorganic pyrophophatases was obtained using a combined strategy of web servers, MODELLER and molecular dynamics simulations. The resulting models provided ground for the lack of quaternary structure in plant pyrophosphatases (Rosales-León et al., 2012). Although the sequences of several related isoforms were initially sent to the servers, only one isoform was correctly modeled, according to Rd.HMM, but the Rd.HMM of the good model gave an alignment for the sequence of a second isozyme. This alignment, and the correct model were then used to produce the second model. Though this last model was not directly based on experimental data, its quality was high, according to Rd.HMM (Rosales-León et al., 2012).

10. Rd-HMMer limitations

Since most Rd.HMM limitations have been mentioned. We only summarize them here:

a. Rd.HMM sensitivity makes it useful for medium to good quality models. Low quality models, may still be of use as starting points, but the Rd.HMM data will only indicate the low quality and will not allow to discriminate a wrong model from an unrefined one.
b. The structurally aware nature of the Rd.HMM alignments is to be trusted only for good quality models. As the Rd.HMM score drops, the sequence to structure correlation becomes weak.
c. Rd.HMM does not offer much information on how to modify the model to improve its appropriateness, other than the presence of insertion/deletions or sequence to structure frame-shifts.
d. A model may be badly refined and get a good Rd.HMM score, as long as Rosetta-design is able to process the backbone coordinates and repack the residues. Therefore, the Rd.HMM score is insufficient information. Information from other software, such as ANOLEA energy (Melo & Feytmans, 1998) or molecular mechanics energy (Hu & Jiang, 2010) is always required to test a model quality.
e. Finally, there is no formal proof for the perfect correspondence between a Rd.HMM high score and the prediction for the 3D-structure of a protein to be native-like. Therefore, from two predictions, of which only one represents the native fold, it might be possible to produce a high Rd.HMM score for the target sequence (a false positive). However, despite our best efforts we have only found the false negative case, *i.e.* a good prediction (or even a 3D-structure from experimental data) may give a low Rd.HMM score. To the best of our knowledge, among the quality assessment methods, this feature is unique to the Rd.HMM protocol.

Chapter 10

Bacterial Promoter Features Description and Their Application on *E. coli in silico* Prediction and Recognition Approaches

Scheila de Avila e Silva and Sergio Echeverrigaray

Additional information is available at the end of the chapter

http://dx.doi.org/10.5772/48149

1. Introduction

The determination of when and how genes are "turned on and off" is a challenge in pos-genomic era. Differences between two species are closer to gene expression and regulation than to gene structures (Howard & Benson, 2002). The first and key step in gene expression is promoter recognition by RNA polymerase enzyme (RNAP). The promoter sequences can be defined as cis-acting elements located upstream of the transcription start site (TSS) of open reading frames (ORF). To make an analogy, genes represent the "computer memory" and promoters represent the "computer program" which acts on that memory. The study about promoters can assist in providing new models about the constitution of the computer program and how it operates (Howard & Benson, 2002).

The proper regulation of transcription is crucial for a single-cell prokaryote since its environment can change dramatically and instantly (Huffmann & Brennan, 2002). In face of this, the detailing of the principals and the organization of transcriptional process is helpful for understanding the complexity of biological systems involved, for instance, cellular responses to environmental changes or in the molecular bases of many diseases caused by microbes (Janga & Collado-Vides, 2007).

While several sequenced genomes have their protein-coding gene repertoire well described, the accurate identification and delineation of cis-regulatory elements remain elusive (Fauteux et al., 2008). At this moment, the challenges are to analyze the available sequences and to locate TSS, promoters and other regulatory sequences (Askary et al., 2009). The purpose of this review is to provide a brief survey of promoter sequences characteristics and the advances of computer algorithms for their analysis and prediction. This chapter is organized in two main sections. The established knowledge about biological features of the

positions for the promoter. A PCSM for promoter and another for non-promoter training sequences sets have been computed. For classifying a new test sequence, the resulted scores from promoter and negative PCSM were used. Based on those scores, the sequence was identified as promoter only if the score was higher for positive PCSM. The results achieved in this paper present sensitivity of 91% and specificity of 81%. In order to predict promoters in the whole genome, the PCSM was applied and all the 683 experimentally identified σ^{70}-dependent promoter sequences were successfully predicted. Besides that, 1567 predictions were considered as probable promoters.

To predict σ^{28} promoter-dependent sequences of ten gamma-proteobacteria species, Song et al. (2007) carried out an alternative approach based on PWM named as Position Specific Score Matrix (PSSM). The species chosen were *E. coli, Bacillus subtilis, Campylobacter jejuni, Helicobacter pylori, Streptomyces coelicolor, Corynebacterium glutamicum, Vibrio cholera, Shewanella oneidensis, Xanthomonas oryzae* and *Xanthomonas campestris*. This approach involved two steps: *(i)* a simple pattern-matching with the short *E. coli* σ^{28} promoter consensus sequence (TAAAG-N$_{14}$-GCCGATAA) for predicting σ^{28} promoters upstream of mobility and chemotaxis genes in test species; *(ii)* these predicted promoters were used to generate a preliminary PSSM for each species. The total length of DNA analyzed for each bacteria was between 4×10^5 bp and 7×10^5 bp. The cut-off values chosen were set to control the false positive rate at 1 every 5×10^5 bp of sequence analyzed using random DNA sequence of 5×10^7 pb. Although the performance measures were not present by the authors, this paper is devoted to predict other promoter sequences than those recognized by σ^{70} and it shows interesting results about the σ^{28} consensual promoter sequences.

PWM models are commonly used because they are a simple predictive approach. Moreover, they are a convenient way to account for the fact that some positions are more conserved, than others (Stormo, 2000). However, in a large number of sequences the consensus can be insufficiently conserved, that is, they present insertions, deletions, variable spacing between elements or they are difficult to define. In such cases, this approach yield many false predictions (Kalate *et al.*, 2003). Another limitation is the assumption that the occurrence of a given nucleotide at a position is independent of the occurrence of nucleotides at other positions (Stormo, 2000). Additionally, the use of this approach is highly influenced by the cut-off value chosen, since low cut-off values encourage a high false positive rate, while high cut-off values encourage a high false-negative rate (Song et al., 2007).

3.3. Machine Learning

Machine Learning (ML) concerns the development of computer algorithms which allow the machine to learn from examples. The classification (or pattern recognition) is an important application of ML techniques in bioinformatics due to their capability of capturing hidden knowledge from data. This is possible to achieve even if the underlying relationships are unknown or hard to describe. Additionally, they can recognize complex patterns in an automatic way or distinguish exemplars based on these patterns (Cen *et al.*, 2010; Sivarao *et al.*, 2010).

ML approaches usually split the data set into training and test groups. They learn from examples (training data), and the set of examples, which were not exposed to the classifier in the training process, are used to test the classification model. Among all ML techniques, Support Vector Machines (SVM) and Artificial Neural Network (ANN) applications have produced promising results in the promoter prediction problem. For this reason, the purpose of this section is to provide an explanation about the basic ideas of these two ML approaches.

3.3.1. Support vector machines

SVM has been applied to identify important biological elements including protein, promoters and TSS, among others. This technique is used in bioinformatics as not only it can represent complex nonlinear functions but it also has flexibility in modeling diverse sources of data. This approach, introduced by Vapnik and his collaborators in 1992, is usually implemented as binary classifiers and it yields results by two key concepts: the separation of the data set into two classes by a hyperplane, and the application of supervised learning algorithms denoted as kernel machines (Ben-Hur et al., 2008; Kapetanovic et al., 2004; Polat & Günes, 2007). In a simple way (Figure 3), SVM classifies the data by: *(i)* drawing a straight line which separates the positive examples in one side and negative examples in the other side and, *(ii)* computing the similarity of two points with the kernel function (Ben-Hur et al., 2008). The kernel function is crucial for SVM, since the knowledge captured from the data set is obtained if a suitable kernel is defined (Ben-Hur et al., 2008). Further information and mathematical background of SVM can be found in Abe (2010), Ben-Hur et al. (2008), and Zhang (2010).

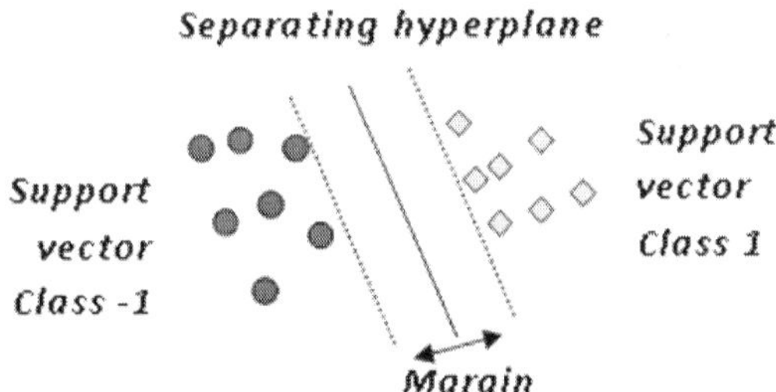

Figure 3. Representation of the basic idea of the SVM classification

Some published paper devoted to promoter prediction using SVM. L. Gordon et al. (2003) carried out SVM with alignment kernel in two different data sets: promoters and coding regions, and promoters and non-promoter intergenic regions. The average error achieved was 16.5% and 18.6%, respectively for the data sets used. This method is preferable in cases which present a sufficient number of known promoter regions, but might not know anything about their composition (L. Gordon et al., 2003). This tool is available online in http://nostradamus.cs.rhul.ac.uk/~leo/sak_demo/. Another SVM carried out by J. J. Gordon

Howard, D.; Benson, K. (2002). Evolutionary computation method for pattern recognition of *cis*-acting sites. *BioSystems*, Vol.72, pp.19-27.

Huerta, A.M.; Collado-Vides, J. (2003) Sigma^{70} promoters in *Escherichia coli*: specific transcription in dense regions of overlapping promoter-like signals, *Journal of Molecular Biology*, Vol.333, pp. 261–278

Huffmann, J. L.; Brennan, R. G. (2002). Prokaryotic transcription regulators: more than just the helix-turn-helix motif. *Current Opinion in Structural Biology*, Vol.12, pp.98-106.

Jacques, P-E.; Rodrigue, S.; Gaudreau, L.; Goulet, J.; Brzezinski, R. (2006). Detection of prokaryotic promoters from the genomic distribution of hexanucleotides pairs. *BMC Bioinformatics*, Vol. 7:423.

Janga, S.C.; Collado-Vides, J. (2007). Structure and evolution of gene regulatory networks in microbial genomes. *Research Microbiology* Vol.158, pp. 787–794.

Jáuregui, R.; Abreu-Goodger, C.; Moreno-Hagelsieb, G., Collado-Vides, J.; Merino, E. (2003) Conservation of DNA curvature signals in regulatory regions of prokaryotic genes. *Nucleic Acids Research*, Vol. 31, No. 23, pp. 6770-6777.

Kalate, R. N., Tambe, S. S., Kulkarni, B. D., Artificial neural networks for prediction of mycobaterial promoter sequences, Computational Biology and Chemistry 27 (2003) 555-564.

Kanhere, A., Bansal, M. (2005a). Structural properties of promoters: similarities and differences between prokaryotes and eukaryotes. *Nucleic Acids Research*, Vol. 33, pp. 3165-3175.

Kanhere, A., Bansal, M. (2005b). A novel method for prokaryotic promoter prediction based on DNA stability. *BMC Bioinformatics*, Vol.6, pp. 1471-2105.

Kapetanovic, M., Rosenfeld, S., Izmirlian, G. (2004). Overview of commonly used bioinformatics methods and their applications. Ann N Y Acad Sci 1020 10–21.

Klaiman, N. T.; Hosid, S.; Bolshoy, A.; (2009). Conservation of DNA curvature signals in regulatory regions of prokaryotic genes. *Computational Biology and Chemistry*, Vol. 33, pp. 275-282.

Kozobay-Avraham, L.; Hosid, S.; Volkovich, Z.; Bolshoy, A. (2006). Involvement of DNA curvature in intergenic regions. *Nucleic Acids Research*, Vol.34, No.8, pp. 2316-2327.

Krishnamachari, A.; Mondal, M.V.; Karmeshu (2004). Study of DNA binding sites using the Renyi parametric entropy measure. *Journal of Theoretical Biology*, Vol. 227, pp. 429–436

Lewin, B., 2008. *Genes IX*. Jones & Bartlett Publishers, ISBN 13:978-0-7637-4063-4, Sudbury.

Li, Q-Z., Lin H. (2006). The recognition and prediction of σ^{70} promoters in Escherichia coli K-12. *Journal of Theoretical Biology*, Vol.242, pp. 135–141.

Lisser, S.; Margalit, H. (1993). Compilation of E. coli mRNA promoter sequences, *Nucleic Acids Research*, Vol.21, No.7, pp. 1507–1516.

Mahadevan, I.; Ghosh, I. (1994). Analysis of *E. coli* promoter structures using neural networks. *Nucleic Acids Research*, Vol.22, pp. 2158-2165.

Nickerson, C.A.;Achberger, E.C. (1995). Role of curved DNA in binding of *Escherichia coli* RNA polymerase to promoters. *Journal of Bacteriology*, Vol.157, pp.5756-5761.

Olivares-Zavaleta, N.; Jáuregui, R.; Merino, E. (2006). Genome analysis of *Escherichi coli* promoters evidences that DNA static curvature plays a more important role in gene transcription than has previously been antecipated. *Genomics*, Vol. 87, pp. 329-337.

Pandey, S. P.; Krishnamachari, A. (2006). Computational analysis of plant RNA Pol-II promoters. *Biosystems*, Vol.83, pp. 38-50.

Pedersen, A. G. & Engelbrecht, J. Investigations of Escherichia coli promoter sequences with artificial neural networks: new signals discovered upstream of the transcriptional startpoint, Proc Int Conf Intell Syst Mol Biol 3 (1995) 292-299

Polat, K.; Günes, S. (2007). A novel approach to estimation of *E. coli* promoter gene sequences: Combining feature selection and least square support vector machine (FS_LSSVN). *Applied Mathematics and Computation*, Vol. 190, pp. 1574-1582.

Potvin, E.; Sanschagrin, F.; Levesque, R. C. (2008). Sigma factors in *Pseudomonas aeruginosa*, *Federation of European Microbiological Societies*, Vol.32, pp. 38–55.

Rangannan, V.; Bansal, M. (2007).Identification and annotation of promoter regions in microbial genome sequences on the basis of DNA stability. *Journal of Biosciences*, Vol.32, No.5, pp. 851-862.

Rani, T.S.; Bhavani, S.D.; Bapi, R.S. (2007) Analysis of *E. coli* promoter recognition problem in dinucleotide feature space. *Bioinformatics*, Vol.23, No.5, pp. 582-588.

Reese, M.G. (2001) Application of a time-delay neural network to promoter annotation in the *Drosophila melanogaster* genome. *Computers and Chemistry*, Vol.26, pp. 51-56.

Rhodius, V. A., W. C. Suh, G. Nonaka, J. West, and C. A. Gross. (2006). Conserved and variable functions of the σ^E stress response in related genomes. PLoS Biol. 4:0043-0059.

Santalucia, J. Jr.; Hicks, D. (2004). The thermodynamics of DNA Structural Motifs. *Annual Review of Biophysics and Biomolecular Structure*, Vol. 33, pp. 415-440.

Shultzaberger, R.K.; Chen, Z.; Lewis, K.A.; Schneider, T.D.; (2007). Anatomy of *Escherichia coli* σ^{70} promoters. *Nucleic Acids Research*, Vol.35, No.3, pp. 771–788.

Sivarao, P. B.; El-Tayeb, N. S. M.; Vengkatesh, V.C. (2010). Neural network multilayer perceptron modelling for surface quality prediction in laser machining, In: *Application of Machine Learning*, Y. Zhang (Ed.), pp.1-20, InTech, ISBN 978-953-307-035-3, India

Sokolova, M.; Lapalme, G. (2009). A systematic analysis of performance measures for classification tasks. *Information Processing and Management*, Vol.427, pp. 427-437.

Song, W.; Maiste, P.J.; Naiman, D.Q.; Ward, M.J. (2007). Sigma 28 promoter prediction in members of the Gammaproteobacteria. *Federation of European Microbiological Societies*, Vol.271, pp. 222-229.

Stormo, G.D. (2000). DNA binding sites: Representation and discovery. *Bioinformatics*, Vol.16, No.1, pp. 16-23.

Thiyagarajan, S.; Rajan, S.; Gautham, N. (2006). Effect of DNA structural flexibility on promoter strength - molecular dynamics studies of *E. coli* promoter sequences. *Biochemical and Biophysical Research Communications*, Vol.341, pp. 557–566.

Typas, A.; Becker, G; Hengge, R. (2007). The molecular basis of selective promoter activation by the σ^S subunit of RNA polymerase. *Molecular Microbiology,* Vol.63, No.5, pp. 1296–1306

Wang, H.; Benham, C. J. (2006). Promoter prediction and annotation of microbial genomes based on DNA sequences and structural responses to superhelical stress. *BMC Bioinformatics*, Vol. 7:248.

WU, Cathy H. (1996). Artificial neural networks for molecular sequence analysis. *Computers & Chemistry*, Vol. 21, No. 4, pp. 237-256.

Zhang, Y. (2010). *New Advances in Machine Learning,* InTech, ISBN 978-953-307-034-6, India.

Chapter 11

Computational Approaches for Designing Efficient and Specific siRNAs

Suman Ghosal, Shaoli Das and Jayprokas Chakrabarti

Additional information is available at the end of the chapter

http://dx.doi.org/10.5772/50125

1. Introduction

Small RNA mediated RNA interference (RNAi) is a widely adopted mechanism towards immunity in plants and invertebrates. Two types of small RNAs- small interfering RNA (siRNA) and microRNA (miRNA), play key role in RNA interference either through cleaving or through translational repression of the target mRNA by guiding RNA induced silencing complex (RISC) to its target site. siRNA is a small RNA (19-23 nucleotides) which is complementary to part of their target mRNA [1]. siRNAs are very efficient in target gene knockdown that makes synthetic siRNA's perfect choice for use in experiments for silencing genes to examine their function. In addition, siRNA have good potential in drug development for therapeutic purpose [2]. Exogenous synthetic siRNAs are designed to target a part of the coding region in the target mRNA [3]. But, as evident from experiments, all siRNAs are not equally efficient in target gene silencing. The potency of siRNAs is largely dependent upon the selection of the region it targets. A displacement of 5-6 nucleotide position hugely alters the efficiency of siRNA. The reason behind this alteration of efficiency is the alteration in local sequence and structural features of the target region. These variations in sequence and structural features correlate with target accessibility, RISC loading or stimulation of immune response. A lot of study has been done in the field of rational siRNA designing to find appropriate parameters that facilitate designing of effective siRNAs [4]. Several commercial suppliers and non-profit educational institutes contribute to the research for searching appropriate siRNA selection parameters for improving potency of designed siRNAs. Progresses have been made in targeting success rate compared to the early days- from as low as 0-10% targeting success rate, today siRNAs have reached average 50% targeting success rate. Still there are many scopes to improve siRNA designing. For efficient designing, the siRNA selection parameters must be arranged and weighted in such a way that ensures optimal result while selecting the siRNA target site. There are some previously suggested guidelines about parameter weight assignment like rational or weighted methods. These guidelines were made in the early days of siRNA selection research with small number of data, when experimentally knockdown validated siRNA dataset was scarce. But with increasing amount of knockdown validated siRNA datasets, new parameter optimization methods must come out to ensure selection of potent siRNAs in a bigger scenario. Today, high throughput siRNA screening experiments have become a common technique to investigate thousands of gene functions at a time to study some specific pathway. These experiments use siRNA libraries targeting transcripts in a genome wide range. But the success of these experiments relies on the knockdown success rate of the siRNAs in the library. After two decades of research, still many of these large scale screening experiments fail because of a large number of non-functional siRNAs present in the libraries used in these experiments. So, still there are needs for improvement in the siRNA selection algorithm and efficient parameter optimization.

Another main challenge in computational siRNA designing lies in the specificity issue of siRNAs. Exogenous siRNAs often induce off-target effects that arise from near perfect or imperfect sequence complementarity with other mRNAs resulting in false positive phenotype during RNAi-based study of gene functions. The type of gene regulation resulting from imperfect sequence complementarity resembles target regulation by endogenous miRNAs- hence this type of off-targeting is called miRNA-like off-targeting. Minimization of miRNA-like off-targets involves choosing a siRNA with a seed region sequence that has fewer targets in the 3′ UTR of other mRNAs. The siRNA seed region is a 6 or 7 nucleotide sequence (the 2-8th nucleotide position) from 5′ end of siRNA guide strand and finding complementary sequence with this 6 or 7 nucleotide seed in the 3′ UTRs of unintended transcripts results in a huge number of potential targets. A majority of these predicted targets are not practically relevant, as in practice, a large number of these predicted off-targets may not be silenced at all. Like miRNA targets, siRNA targets are also dependent on target accessibility and other sequence features around the target site. So, a more rational approach is needed when predicting the siRNA off-targets resulting from partial sequence complementarity. This chapter focuses on current siRNA designing parameters and the approaches towards minimizing the off-target effects as provided by the in-silico siRNA designing solutions and then discuss a bit about a customized off-target reducing algorithm that will be useful to avoid particular genes from being off-targeted.

2. Guidelines for selecting potent siRNAs

As discussed earlier, potency of siRNAs varies greatly with selection of target region. The parameters responsible for effectiveness of siRNAs being target accessibility, uniqueness of target region, absence of SNP sites etc. Other parameters worth considering while selecting the target site is the consideration of alternatively spliced isoforms of the target gene [5].

A parameter which greatly influences siRNA potency is RISC loading of the intended antisense strand [6]. Only one strand of the siRNA duplex enters into RISC and guides the complex to the target mRNA for its silencing. So, the choice of strand that enters into RISC complex plays the vital role in the whole silencing process. Generally RISC complex chooses the siRNA strand which has weaker 5′ end binding energy. So this thermodynamic property should be ensured while selecting siRNAs. Also there are a number of sequence compositions that are said to be important for recognition of the intended strand by RISC and some parameters for enhancing the stability of siRNA duplex are also worth considering while selecting siRNAs. These rules can be divided into three categories 1) Selection of target region, 2) Structural and Thermodynamic Consideration and 3) Sequence Characteristics.

2.1. Selection of target region

Selection of the target region for siRNA should satisfy some constraints to ensure effective silencing. Target site should be analyzed for its position within the mRNA, presence of polymorphic site or other considerations like homology with other mRNAs or its alternatively spliced isoforms.

2.1.1. Selection of target starting site

Generally siRNAs are targeted to a part of the coding region of an mRNA. The target site should be deep inside the open reading frame (ORF) of the target mRNA for efficient silencing. Generally it is advised to start from 50-100 nucleotides downstream of the AUG start codon. Besides targeting the coding region, some siRNAs may also be designed to target the 3′ un-translated region (UTR) of an mRNA. This type of targeting is especially done in case of experiments conducted for restoration of original phenotype. In such cases it is advised to start targeting regions at least 15-20 nucleotides downstream of the stop codon.

2.1.2. Consideration for alternative splicing

In eukaryotic organisms frequent alternative splicing results in diversification of mRNAs. To account for the alternative splicing, it is necessary to evaluate a common target region where siRNAs can be designed to knockdown a specific mRNA isoform or multiple mRNA isoforms from a target gene.

2.1.3. Absence of homology with other mRNAs

Cellular mRNAs with 15/16 or more consecutive base match with the siRNA, are likely to be silenced and degraded by the siRNA. So target sites with 15 or more consecutive base homology with any other mRNA should be avoided.

2.1.4. Avoiding target sites having polymorphic locus

Target regions having single nucleotide polymorphic (SNP) sites should be avoided.

2.2. Structural and thermodynamic consideration

siRNA potency largely depends upon structural constraints of the target region. Heavily structured sites are less likely to be bound by siRNAs as these sites are not accessible by siRNAs. The relative binding energy of the 5′ and 3′ ends of the siRNA with the target site play a vital role in the choice of strand to be incorporated into RISC complex and thus is one of the most important parameter to be considered during siRNA design.

2.2.1. Presence of Secondary structure

It has been suggested that presence of local secondary structures (stem loops) in the target site restricts its accessibility to RISC and hence reduces the efficiency of the siRNA. So it is necessary to filter out those potential inaccessible target sites with strong secondary structures. The prediction of local secondary structure can be made by numerous RNA secondary structure prediction tools or packages like Mfold [7] or Vienna RNA package [8] - that mainly predict minimum free energy secondary structure of a RNA sequence.

2.2.2. Thermodynamic property for efficient RISC loading

In a siRNA duplex, antisense strand with relatively low energy in 5′ end is favourable for its loading into RISC complex. So, there should be difference in binding energy between the 5′ end of the sense and antisense strand.

2.3. Sequence characteristics

Years of research for finding appropriate designing parameters identified some sequence parameters enriched within efficient siRNAs. These sequence characteristics often contribute to efficient RISC loading or siRNA sequence specificity or stability issues.

2.3.1. Position specific nucleotide composition

Sequence analysis of effective siRNAs revealed many position specific nucleotide compositions for enhancing potency of the siRNA. Some of these preferences are listed below in table 1-

2.3.2. Sequence feature for efficient RISC entry

siRNA guide strands with low energy at 5′ end are favored for entering the RISC complex. So, presence of at least three (A/U)s in the seven nucleotides at the 3' end of the sense strand is preferable.

2.3.3. siRNA duplex stability

Target sites with low GC content (generally less than 55%) has a greater potential for being functional siRNA site, as too high GC content can impede the loading of siRNAs into RISC complex. Also, too low GC content is not favourable because too low GC content can destabilize the siRNA duplex and reduce their affinity to target mRNA binding. Analysis of effective siRNAs showed G/C content between 35% and 60% is most favorable.

Position specific nucleotides
Presence of A base at position 19 of the sense strand.
Presence of U base at position 10 of the sense strand.
Presence of A base at position 3 of the sense strand.
A base other than G or C at position 19 of the sense strand.
A base other than G at position 13 of the sense strand.
Presence of A base at the 2nd nucleotide position of the sense strand.
Presence of C base at the 4th nucleotide position of the sense strand.
Absence of C base at the 6th nucleotide position of the sense strand.
Absence of U base at the 7th nucleotide position of the sense strand.
Presence of C base at the 9th nucleotide position of the sense strand.
Presence of A base at the 17th nucleotide position of the sense strand.
Absence of C base at the 18th nucleotide position of the sense strand.
No occurrences of four or more identical nucleotides in a row.
No occurrences of G/C stretch of length 7 or longer.

Table 1. Position specific nucleotide composition prefered in functional siRNAs

3. Choice of appropriate parameters

All the parameters discussed above are not equally important for selection of efficient siRNAs. By far, many research groups have conducted studies for evaluation of effective parameter sets for siRNA selection. Gong et al. studied 276 known siRNA selection parameters on a sufficiently large set of 3277 experimentally validated siRNAs targeting 1518 genes to identify common parameters that effectively distinguishes functional siRNAs from non functional ones [9]. They were able to identify 34 features associated with improved siRNA efficacy among which 27 features were associated with greater than 70% efficacy. They examined combination of siRNA features to find their cooperative effects on potent siRNA selection and used a disjunctive rule merging (DRM) algorithm to generate a bunch of non-redundant rules set to efficiently predict functional siRNAs and lower the false positive predictions. Table 2 list 17 features set associated with greater than 90% efficacy and used for optimal features combination.

[12] Ren Y., Gong W., Xu Q., Zheng X., Lin D., Wang Y., Li T. (2006). siRecords: an extensive database of mammalian siRNAs with efficacy ratings. *Bioinformatics,* Vol. 22, (January 2006), pp. (1027)

[13] Platt J. C. (1999) Fast training of support vector machines using sequential minimal optimization, Advances in kernel methods. *MIT Press Cambridge,* pp. (185)

[14] MIT siRNA database [http://web.mit.edu/sirna/sirnas-human.html].

[15] Jackson A.L. & Linsley P.S. (2010). Recognizing and avoiding siRNA off-target effects for target identification and therapeutic application. *Nature Review Drug Discovery,* Vol. 9, (January 2010), pp. (57)

[16] Judge D., Sood V., Shaw J.R., Fang D., McClintock K., MacLachlan I. (2005). Sequence-dependent stimulation of the mammalian innate immune response by synthetic siRNA. *Nature Biotechnology,* Vol. 23, (March 2005), pp. (457)

[17] Fedorov Y., Anderson E.M., Birmingham A., Reynolds A., Karpilow J., Robinson K., Leake D., Marshall W.S., Khvorova A. (2006). Off-target effects by siRNA can induce toxic phenotype. *RNA,* Vol. 12, (March 2006), pp. (1188)

[18] Birmingham A., Anderson E.M., Reynolds A., Ilsley-Tyree D., Leake D., Fedorov Y., Baskerville S., Maksimova E., Robinson K., Karpilow J., Marshall W.S., Khvorova A. (2006). 3' UTR seed matches, but not overall identity, are associated with RNAi off-targets. *Nature Methods,* Vol. 3, (March 2006), pp. (199)

[19] Burchard J., Jackson A.L., Malkov V., Needham R.H.V., Tan Y., Bartz S.R., Dai H., Sachs A.B., Linsley P.S. (2009). MicroRNA-like off-target transcript regulation by siRNAs is species specific. *RNA,* Vol. 15, (February 2009), pp. (308)

[20] Wang L. & Forest Y.M. (2004). A Web-based design center for vector-based siRNA and siRNA cassette. *Bioinformatics,* Vol. 20, (September 2004), pp. (1818)

[21] Nielsen C.B., Shomron N., Sandberg R., Hornstein E., Kitzman J., Burge C.B. (2007). Determinants of targeting by endogenous and exogenous microRNAs and siRNAs. *RNA,* Vol. 13, (November 2007), pp. (1894)

[22] Ui-Tei K., Naito Y., Nishi K., Juni A., Saigo K. (2008). Thermodynamic stability and Watson–Crick base pairing in the seed duplex are major determinants of the efficiency of the siRNA-based off-target effect. *Nucleic Acids Research,* Vol. 36, (November 2008), pp. (7100)

[23] Yuan B., Latek R., Hossbach M., Tuschl T., Lewitter F. (2004). siRNA Selection Server: an automated siRNA oligonucleotide prediction server. *Nucleic Acids Research,* Vol. 32, (July 2004), pp. (130)

[24] Anderson E. M., Birmingham A., Baskerville S., Reynolds A., Maksimova E., Leake D., Fedorov Y., Karpilow J. & Khvorova A. (2008). Experimental validation of the importance of seed complement frequency to siRNA specificity. *RNA,* Vol. 14, (May 2008), pp. (853)

[25] Kiryu H., Terai G., Imamura O., Yoneyama H., Suzuki K., Asai K. (2011). A detailed investigation of accessibilities around target sites of siRNAs and miRNAs. *Bioinformatics,* Vol. 27, (July 2011), pp. (1788)

[26] Schultz N., Marenstein D.R., De Angelis D.A., Wang W.Q., Nelander S., Jacobsen A., Marks D.S., Massagué J., Sander C. (2011). Off-target effects dominate a large-scale RNAi screen for modulators of the TGF-β pathway and reveal microRNA regulation of TGFBR2. *Silence,* Vol. 2, (March 2011)

Chapter 12

Novel microRNA Cloning Using Bioinformatics

Yoshiaki Mizuguchi, Takuya Mishima, Eiji Uchida and Toshihiro Takizawa

Additional information is available at the end of the chapter

http://dx.doi.org/10.5772/53945

1. Introduction

MicroRNAs (miRNAs) participate in several biological processes, including development, differentiation, apoptosis, and proliferation (**1, 2**) through imperfect pairing with target messenger RNAs (mRNAs) of protein-coding genes and transcriptional or post-transcriptional regulation of their expression (**3, 4**). Approaches to miRNA detection, such as parallel sequencing technologies may replace conventional sequencing (**5**). The GS 454 technology can produce a similar number of longer (100–150-nucleotides (nt)) sequence reads in a single analysis run, with the advantage that this method can derive the complete sequence of the mature miRNA. Moreover, recent studies on miRNA profiling performed with cloning techniques suggest that sequencing methods are suitable for the detection of novel miRNAs, modifications, and precise compositions, and that cloning frequencies calculated by clone count analysis strongly correlate with the concentrations measured by Northern blotting, and are reproducible. The achievement of comprehensive profiling of miRNA in human diseases requires exhaustive qualitative and quantitative analyses. Here we show the techniques and the some of the results of the miRNA transcriptomes in the liver using sequencing. This serves as a critical step in clarifying the functional significance of specific miRNAs as they relate to liver diseases.

2. Techiniques of MicroRNA cloning and bioinformatics for MiRNAome

2.1. MicroRNA cloning

The method for microRNA cloning and sequencing that we moderated from the original ones are shown in Fig1. We cloned small RNA by a modification of the published miRNA cloning protocol of Lagos-Quintana et al. (**6**). In brief, total RNA samples were extracted using ISOGEN (Nippon Gene, Tokyo, Japan), separated in a denaturing polyacrylamide gel, and the 18–24 nt fraction was recovered. Next, 5′- and 3′-adapters were ligated to the RNAs Ligation of small RNAs with DNA_RNA chimera linkers at both termini [3′ linker

NONCODE, http://www.bioinfo.org.cn/NONCODE/; NCBI Reference Sequence, ftp://ftp.ncbi.nih.gov/refseq/; UCSC Genome Bioinformatics Site, http://genome.ucsc.edu; OncoDb HCC, http://oncodb.hcc.ibms.sinica.edu.tw/index.ht

Chapter 13

Ensemble Clustering for Biological Datasets

Harun Pirim and Şadi Evren Şeker

Additional information is available at the end of the chapter

http://dx.doi.org/10.5772/49956

1. Introduction

Recent technologies and tools generated excessive data in bioinformatics domain. For example, microarrays measure expression levels of ten thousands of genes simultaneously in a single chip. Measurements involve relative expression values of each gene through an image processing task.

Biological data requires both low and high level analysis to reveal significant information that will shed light into biological facts such as disease prediction, annotation of a gene function and guide new experiments. In that sense, researchers are seeking for the effect of a treatment or time course change befalling. For example, they may design a microarray experiment treating a biological organism with a chemical substance and observe gene expression values comparing with expression value before treatment. This treatment or change make researchers focus on groups of genes, other biological molecules that have significant relationships with each other under similar conditions. For instance, gene class labels are usually unknown, since there is a little information available about the data. Hence, data analysis using an unsupervised learning technique is required. Clustering is an unsupervised learning technique used in diverse domains including bioinformatics. Clustering assigns objects into the same cluster, based on a cluster definition. A cluster definition or criterion is the similarity between the objects. The idea is that one needs to find the most important cliques among many from the data. Therefore, clustering is widely used to obtain biologically meaningful partitions. However, there is no best clustering approach for the problem on hand and clustering algorithms are biased towards certain criteria. In other words, a particular clustering approach has its own objective and assumptions about the data.

Diversity of clustering algorithms can benefit from merging partitions generated individually. Ensemble clustering provides a framework to merge individual partitions from different clustering algorithms. Ensemble clustering may generate more accurate clusters than individual clustering approaches. Here, an ensemble clustering framework is implemented as described in [10] to aggregate results from K-means, hiearchical clustering and C-means algorithms. We employ C-means instead of spectral clustering in [10]. We also use different

data sets. Two different biological datasets are used for each algorithm. A comparison of the results is presented. In order to evaluate the performance of the ensemble clustering approach, one internal and one external cluster validation indices are used. Silhouette (S) [31] is the internal validation index and C-rand [23] is the external one. The chapter reviews some clustering algorithms, ensemble clustering methods, includes implementation, and conclusion sections.

2. Clustering algorithms

Clustering biological data is very important for identification of co-expressed genes, which facilitates functional annotation and the elucidation of biological pathways. Accurate predictions can serve as a guide for targeting further experiments and generating additional hypotheses. Furthermore, accurate predictions can facilitate identification of disease markers and targets for drug design [4]; clustering can also be used to determine whether certain patterns exist near viral integration sites[16].

Current algorithms used in gene clustering have some drawbacks. For example, K-means algorithm is sensitive to noise that is inherent in gene expression data. In addition, the solution (i.e. the final clustering) that the K-means algorithm finds may not be a global optimum since it relies on randomly chosen initial objects. However, K-means-based methods are prevalent in the literature such as [12, 17, 33]. K-means works upon randomly chosen centroid points that represent the clusters. The objects are assigned to the closest clusters based on distance calculation regarding centroid points. For example, the dataset illustrated in Figure 1 is assigned two centroids.

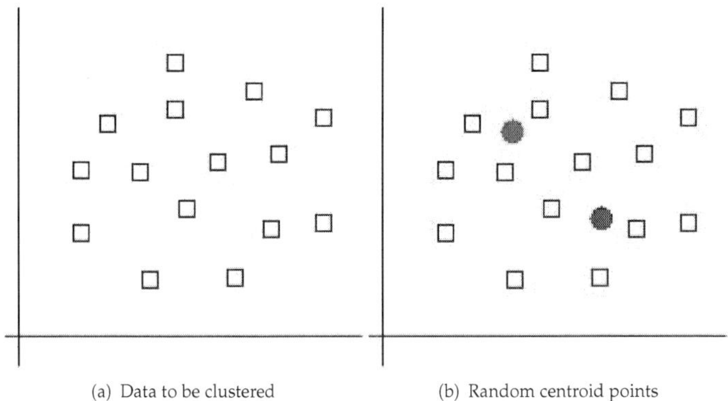

(a) Data to be clustered   (b) Random centroid points

Figure 1. The dataset and two centroid points

The distance between any object from the dataset to both of the centroid points are calculated and the objects are assigned to the closest cluster represented by the closest centroid point as seen in Figure 2. Then new centroid points of clusters are calculated and objects are assigned to the closest clusters regarding the distance to new centroid points. Recalculation of centroid points and assignment of objects to new clusters goes on till centroids points remain the same as in Figure 3.

Another method, Self-organizing Map (SOM), is one of the machine-learning techniques widely used in gene clustering. A recent study is [14]. SOM requires a grid structured input that makes it ineffective.

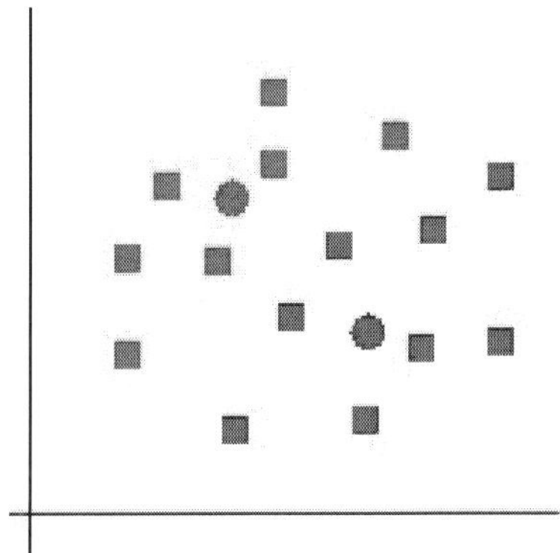

Figure 2. Initial clusters

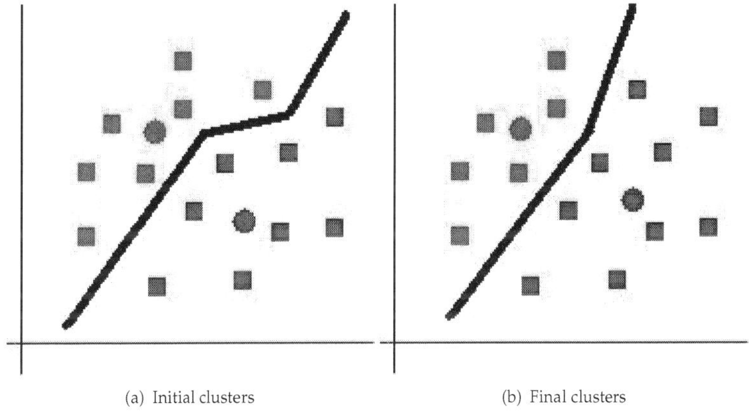

(a) Initial clusters   (b) Final clusters

Figure 3. Iteration of K-means

Hierarchical clustering (HC) algorithms are also widely used and area of two types: agglomerative and divisive. In agglomerative approach objects are all in different clusters and they are merged till they are all in the same cluster as seen in Figure 4.   Two important drawbacks of the HC algorithms are that they are not robust and they have high computational complexity.   HC algorithms are "greedy" which often means that the final solution is suboptimal due to locally optimal choices being made in initial steps, which turn out to be poor choices with respect to the global solution. A recent study is [26].

Graph-theoretical clustering techniques exist in which the genomic data are represented by nodes and edges of a graph.   Network methods have been applied to identify and characterize various biological interactions [13]. Identification of clusters using networks is

fuzzy t-norms algorithm, and finally, the fuzzy K-means algorithm is applied to the rows of the similarity matrix to obtain the consensus clustering. It is demonstrated that the proposed ensemble approach is competitive with the other ensemble methods.

High throughput data may be generated by microarray experiments. If the dataset is very large, it is possible to generate an ensemble of clustering solutions, or partition the data so that clustering may be performed on tractable-sized disjoint subsets [20]. The data can then be distributed at different sites, for which a distributed clustering solution with a final merging of partitions is a natural fit. [20] introduce two new approaches to combining partitions represented by sets of cluster centers. It is stated that these approaches provide a final partition of data that is comparable to the best existing approaches and that the approaches can be 100,000 times faster while using much less memory. The new algorithms are compared with the best existing cluster ensemble approaches that cluster all of the data at once, and a clustering algorithm designed for very large datasets. Fuzzy and hard K-means based clustering algorithms are used for the comparison. It is demonstrated that the centroid-based ensemble merging algorithms presented in the study generated partitions which are as good as the best label vector method, or the method of clustering all the data at once. The proposed algorithms are also more efficient in terms of speed.

[11] propose evidence accumulation clustering based on dual rooted prim tree cuts (EAC-DC). The proposed algorithm computes the co-association matrix based on a forward algorithm that repeatedly adds edges to Prim's minimum spanning tree (MST) to identify clusters until a satisfying criterion is met. A consensus cluster is then generated from the co-association matrix using spectral partitioning. Here, a MST is a fully connected sub-graph with no cycles and a dual-rooted tree is obtained by finding the union of two sub-trees. They test their approach using the Iris dataset [8], the Wisconsin breast cancer dataset [27] (both obtained from [9]) and synthetic datasets, and presented a comparison of their results with other existing ensemble clustering methods.

[22] use a cluster ensemble in gene expression analysis. In the proposed ensemble framework, the partitions generated by each individual clustering algorithm are converted into a distance matrix. The distance matrices are then combined to construct a weighted graph. A graph partitioning approach is then used to generate the final set of clusters. It is reported that the ensemble approach yields better results than the best individual approach on both synthetic and yeast gene expression datasets.

[10] merge multiple partitions using evidence accumulation. Each partition generated by a clustering algorithm is used as a new piece of knowledge, to help uncover the relationships between objects. For this chapter, we adopt their ensemble approach. The core idea behind the ensemble approach here is constructing the co-association matrix by employing a voting mechanism for the partitions generated using individual clustering algorithms. A co-association matrix C is constructed based upon the formulation below, where n_{ij} is the number of times the object pair (i,j) is assigned to the same cluster among the N different partitions:

$$C(i,j) = \frac{n_{ij}}{N}$$

After constructing the co-association matrix, [10] use single linkage hierarchical clustering to obtain the new cluster tree (dendrogram) and then use a cut-off value corresponding to the maximum life time (difference between merge points where branching starts) on the tree.

They also employ the same ensemble framework using K-means partitions with different parameters. They test their algorithms on ten different datasets, comparing the results with other ensemble clustering methods. They report that their ensemble approach can identify the clusters with arbitrary shapes and sizes, and perform better than the other combination methods.

4. Implementation

We employ the ensemble approach described in [10].   Different set of base clustering algorithms are chosen and implemented on protein and lymphoma datasets.

Protein dataset consists of 698 objects (corresponding to protein folds) with 125 attributes. The protein dataset contains 698 proteins from 125 samples. The real clusters correspond to the four classes of protein−folds: α, β, α/β and $\alpha+\beta$ protein classes. DLBCL−B is 2−channel custom cDNA microarray dataset. This is a B cell lymphoma dataset with predefined three subtypes [21].

The ensemble clustering algorithm uses an array of vectors data structure for each of the file, in order to use the dynamic memory allocation and starts with initializing the file content in the vectors. The algorithm also processes the vectors and generates two temporary matrices with the dimension of maximum vector length. The ensemble clustering algorithm steps are as follows:

Algorithm 1 Ensemble Clustering Algorithm

Require: partitions
Ensure: distance matrix

  for $i = 0$ *to* $max(V[n])$ **do**
    for $j = 0$ *to* $max(V[n])$ **do**
      for $k = 0$ *to* n **do**
        if $V[k].elementAt(i) = V[k].elementAt(j))$ **then**
          $C[i][j] = C[i][j] + 1/n$
        end if
        $D[i][j] = 1 - C[i][j]$
      end for
    end for
  end for

Here, n is the number of files, $V[n]$ are the vectors holding the content of each file. $max(V[n])$ is the length of the longest vector, $C[i][j]$ is the co-association matrix and $D[i][j]$ is the distance matrix. The algorithm iterates through the two dimensional matrix via i and j loop variables inside a nested loop at lines 1 and 2 and for each member of the matrix, all the vectors are processed inside the loop via k loop variable at line 3. The condition of equality for the selected vector with the selected loop variables i and j, causes an increase on the co-association matrix elements at lines 4 and 5. Finally the distance matrix is calculated at line 7. After obtaining the distance matrix, hierarchial clustering with complete linkage is used to generate the dengrogram. The dendrogram is cut at a certain level to obtain consensus partition.

Ensemble approach is coded as a java application which is available upon request. The software allows addition of many partitions to generate the distance matrix of the corresponding ensemble. Files including the partitions can be added by clicking on the "Add File" button as seen in Figure 6. Distance matrix of the ensemble is generated by "Calculate" button.

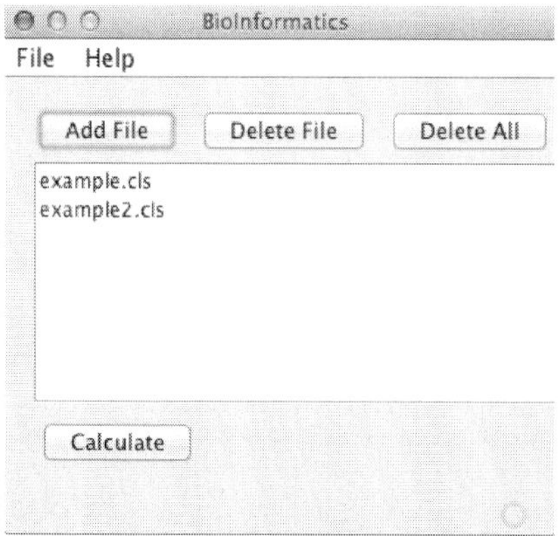

Figure 6. File input interface

The output is displayed on a separate screen as demonstrated on Figure 7. The output with csv format can be written into a file by clicking on the "Output CSV" button.

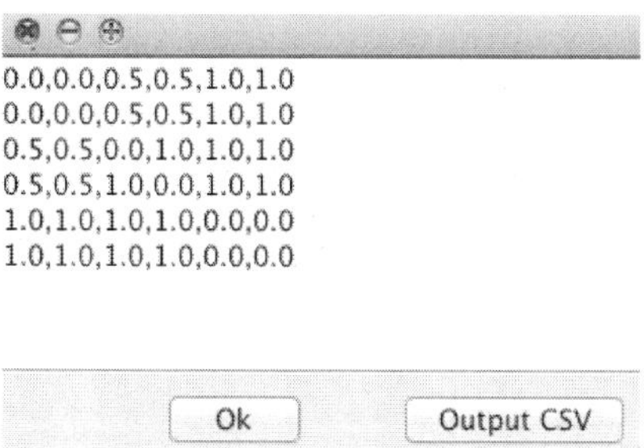

Figure 7. Example clusters

Considering two different partitions of a dataset with six objects which are $(1, 1, 2, 1, 3, 3)$ and $(2, 2, 2, 1, 3, 3)$, the algorithm's output is the distance matrix:

$$\begin{pmatrix} 0 & 0 & 0.5 & 0.5 & 1 & 1 \\ 0 & 0 & 0.5 & 0.5 & 1 & 1 \\ 0.5 & 0.5 & 0 & 1 & 1 & 1 \\ 0.5 & 0.5 & 1 & 0 & 1 & 1 \\ 1 & 1 & 1 & 1 & 0 & 0 \\ 1 & 1 & 1 & 1 & 0 & 0 \end{pmatrix}$$

The distance matrix is used in hierarchical clustering with complete linkage and the following dendrogram is generated. The dendrogram is cut at a level to give three clusters. The corresponding partition is (1, 1, 1, 2, 3, 3) which is the same as second partition (2, 2, 2, 1, 3, 3).

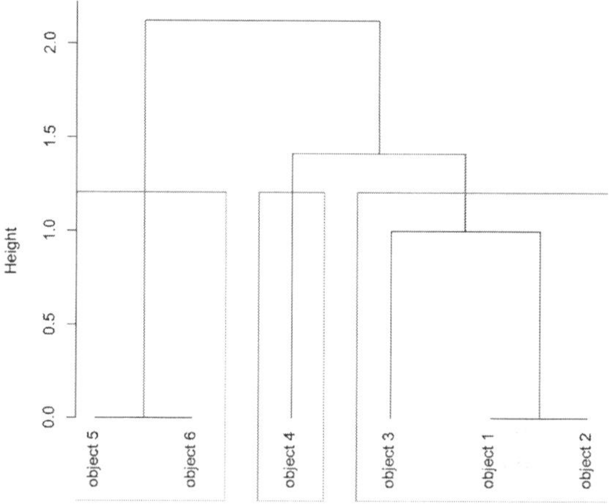

Figure 8. Example clusters

We employ hierarchical clustering, K-means and C-means to obtain base partitions. K-means and hierarchical clustering algorithm are implemented using R base package, C-means is implemented using R e1071 package. Silhouette and C-rand indices are utilized to evaluate the performance of individual and ensemble algorithms. Silhouette and C-rand values are calculated using R clusterSim and flexclust packages respectively. Silhouette is an internal measure of compactness and separation of clusters [6]. The silhouette index values are between -1 and 1 representing worst and best values. C-rand is an external measure of agreement between two partitions. C-rand has maximum value of 1 and it can take negative values. The silhouette and C-rand values found by the base and ensemble algorithms are given in Table 1. Ensemble approach improves clustering result both for the protein and DLBCL-B datasets. Ensemble approach finds better C-rand value, 0.157 than values by K-means and C-means, 0.127 for the protein dataset. Ensemble approach also finds the best C-rand value, 0.135 compared to values generated by individual clustering algorithms, 0.021, 0.063, 0.098. However, the ensemble approach makes S values worse in most cases.

Dataset	Method	Num. of clusters	S value	C value
Protein	HC	4	0.344	0.199
	K-means	4	0.379	0.127
	C-means	4	0.379	0.127
	Ensemble	4	0.078	0.157
DLBCL-B	HC	3	-0.034	0.021
	K-means	3	-0.015	0.063
	C-means	2	-0.005	0.098
	Ensemble	3	-0.017	0.135

Table 1. Index values for base and ensemble algorithms

3. Algorithms for PMWL

The results of traditional matching algorithm are complete, so the focus of research is to improve the matching efficiency. As a kind of searching problem, the key to solving matching problem is how to use and extract information getting from text and pattern. KMP, BM algorithm uses automata to describe the pattern characteristics, and deposit information obtained from scanning during matching process into automata. Algorithm visits the automata, when the jump distance needs to be calculated, thus to avoid obtaining the pattern information repeatedly and to ensure the jump in matching process does not affect the final result. The basic idea the suffix tree is to use the tree structure to describe the text information, and to avoid scanning the same text repeatedly when matching a set of patterns. We believe that data structure and search strategy are crucial for traditional algorithms to access to information of text and pattern. Reasonable data structure is better to explore the potential of the computer, such as bit parallel technology, and can also be a more reasonable representation of the sequence information, such as automata. In addition, there exist the sliding window, indexes and other data structures. Reasonable matching strategy makes better use of sequence information. These strategies can approximately be divided into prefix searching, suffix searching and factor searching (Navarro & Raffinot, 2001).

	Characteristics	Representative algorithms	Methods	Remarks
prefix searching	Forward search to find the longest common prefix of text and pattern strings in searching window	KMP Shift-And	Deterministic automata Bit parallel, non-deterministic automata	Most of them are sliding window technique, the scope of algorithm application depends on the alphabet size and the pattern length
suffix searching	Backward search to find the longest common suffix of text and pattern strings, can skip some text characters, the difficulty is how to safely move the window	BM Horspool	Pre-calculation of the three functions used to determine the safe jumping distance Improve the function of the BM, can have a greater jump distance, especially suitable for larger alphabet	
factor searching	Backward search to determine whether the suffix of text in searching window is a substring of pattern string	BDM BNDM BOM	Suffix automaton Bit parallel Factor Oracle automaton, suitable for longer pattern	

Table 1. Analysis of the traditional pattern matching algorithms

As different extension of traditional matching problem, PMWL problem, approximate matching, and swap matching all belong to the *Non-standard Stringology* problem. Problems in this field mostly belong to the optimization problem, and most of them have not yet been completely solved, such as PMWL and approximate matching problems with wildcards etc. What PMWL and traditional matching problem have in common are:

1. From the view of algorithm itself, the data structures and matching strategies are as the key of algorithm design.
2. From the view of describing the object, how to effectively describe the patterns and text information is the key to solve the problem.

What PMWL and traditional matching problem have in difference are:

1. For PMWL, there is no complete solving yet, so algorithm evaluation criteria include both time efficiency and solution quality; but traditional matching algorithm is only concernèd with matching time.
2. The flexibility and complexity of the PMWL problem definition are reflected in the pattern, therefore, compared with traditional matching, PMWL pay more attention to the description of the pattern information. Pattern characteristics are extremely associated with the solution of PMWL problem.

Next, we will give the representative algorithms for solving PMWL problem, and detailed description of their design ideas from the perspective of data structure and matching strategy.

3.1. The SAIL Algorithm

Description of SAIL Algorithm (Chen et al., 2006):

Input: A text $T = t_0t_1...t_{n-1}$, a pattern $P = p_0p_1...p_{m-1}$, local constraints $g_i = g(N_i, M_i)$, global constraints $[minLen, maxLen]$.

Output: Occurrences of P in T satisfying the constraints.

The Steps of the algorithm:

1. *Location*: ① Search position i where $t[i] = p[m-1]$, and locate position k where $t[k] = p[0]$ by considering the global constraint. ② Cut out a substring in T from $t[k]$ to $t[i]$ named T'. ③ Build the table with the row and column according to T' and P.
2. *Forward*: Scan the table forward, and mark all the positions satisfying the local and global constraints. They are the potential matching positions.
3. *Backward*: Scan the table backward, and select the *left-most* position in the marked cells every row that compose an occurrence. Then mark them used.

Generally, SAIL starts from the beginning of T to search position i where $t[i] = p[m-1]$. After that, SAIL conducts two phases, the *Forward* phase and the *Backward* phase. In the *Forward* phase, SAIL determines whether there is a potential matching occurrence by using a search table. Afterwards, if a potential matching occurrence can be determined, *Backward* phase is triggered out to output an optimal occurrence by using the *left-most* strategy.

3.2. The RSAIL Algorithm

Description of RSAIL Algorithm (Wang et al., 2010):

Definition 6 Given a pattern P, if there are letters $p[i] = p[j]$ where $0 \le i \le m\text{-}1, 0 \le j \le m\text{-}1$, P is called a **pattern with Recurring characters,** and **R pattern** in short, such as a¢[0,1]c¢[0,1]c¢[0,1]t.

Definition 7 Given a pattern P, if all the letters in P are different, P is called a **pattern with No-Recurring characters** and **NR Pattern** for brevity, such as a¢[0,1]c¢[0,1]g¢[0,1]t.

Definition 8 Given a pattern P, if there is a position i such that $p[i] = p[i+1] = \ldots\ldots = p[m\text{-}1]$ where $1 \le i < m\text{-}1$, P is called a pattern with recurring tail characters and **RT pattern** in short. Such as a¢[0,1]c¢[0,1]c. As we can see, the RT pattern is a special form of the R pattern.

From the above discussion, in the research of Chen et al., since they only concern about the on-line situation, their proof of SAIL's completeness is incomplete, which is only suitable for the on-line situation. What is more, it ignores the interaction between different occurrences.

We find that SAIL satisfies the completeness under a certain restriction, i.e. the pattern with no-recurring character (NR pattern), such as a¢[0, 1]t¢[0, 1]g¢[0, 1]c¢[0, 1]. The concept of NR pattern has practical significance, for example, in text mining, where the text is a sequence of words, the NR pattern reflects the semantic relation between words.

We utilize the symmetry to scan the text and the pattern. Then convert an RT pattern into an R pattern.

1. According to the characteristic of P, if it is not an RT pattern, we directly call SAIL; otherwise go to (2);
2. Reverse T and P, respectively get T', P';
3. Call SAIL, and obtain the occurrences of P' in T';
4. Obtain the occurrences of P in T by coordinate transformation of the obtained solution.

Obviously, since the time of the identification of pattern's characteristics is linear, O (RSAIL) = O (SAIL).

Experiments and Analysis:

We will give a set of experiments to illustrate two problems:

1. Analysis of the complete extent of the SAIL algorithm;
2. The comparison of the complete extent of RSAIL and SAIL algorithm;

Considering there is no algorithm can obtain the completeness occurrences of PMWL problem in polynomial time, we have developed a text generator (Xie et al., 2010) to generate experimental text, by which way we can know the completeness occurrences in order to analyze the complete extent of algorithm. In addition, the patterns used in these experiments are all RT patterns.

	Experiment1	Experiment 2	Experiment 3	Experiment 4
Size of alphabet Σ	4	4	4	7
Length of pattern m	3	4	5	5
gap	0~30	0~30	0~30	0~30

Table 5. The parameters in the experiments.

The experimental results and analysis:

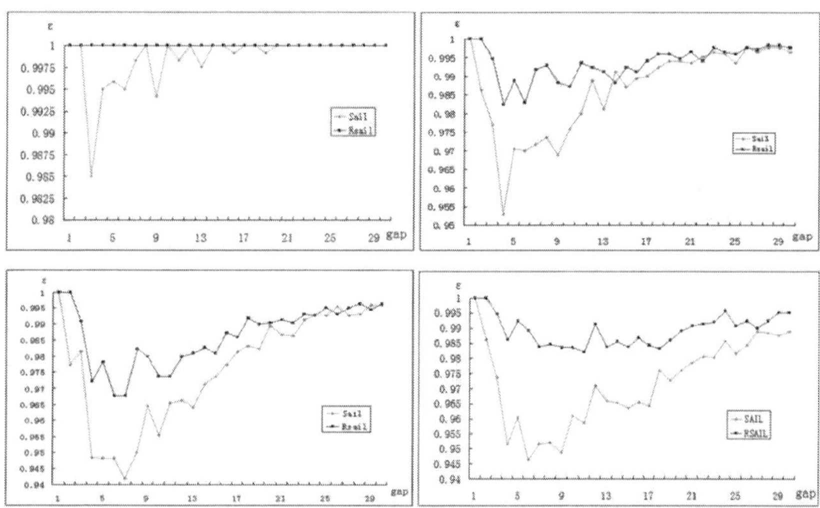

Figure 3. The approximate ratio experimental results of RSAIL and SAIL

From the above images, SAIL itself is already a near-complete algorithm, in the above graphs, the average approximation ratio of SAIL is higher than 0.94;For the RT patterns, the completeness of RSAIL is better than SAIL in different Σ, m and *gap*. Thus, not only a revised algorithm is obtained, the SAIL's deficiency on handling RT patterns is also proved from another aspect.

3.3. The BPBM Algorithm

Description of BPBM Algorithm (Guo et al., 2011):

Like the SAIL algorithm, the BPBM algorithm also focuses on pattern matching in online sequential text with both flexible *gap* constraints by user's specification and the *one-off* condition. BPBM is based on bit-parallel technology to simulate the matching process and adopt two nondeterministic finite state automatons (NFAs). One is a search mechanism to identify all pattern P's suffix, and another one is a security window transition mechanism which accelerates the scanning process by dropping useless sequences in text.

After a series of experiments, we speculate that $\varepsilon = A*gap^4 + B*gap^3 + C*gap^2 + D*gap + E$, where A, B, C, D and E are parameters and for different m there are different parameters. We try to use this model to illustrate the relation between gap and approximation ratio ε.

Use this parameter table, some of illustrations for $m = 3, 4\ldots\ldots14$ are listed below, where horizontal axis is the gap, vertical axis is the ε.

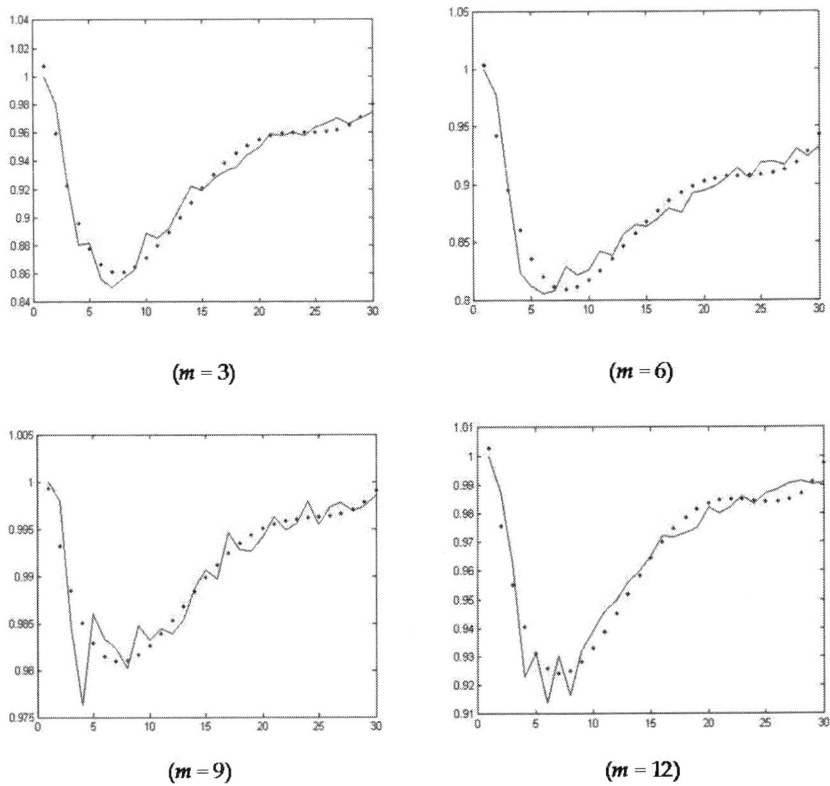

Figure 6. Model fitting

We believe this model can be used to predict the completeness of solutions given a certain pattern. For example, given $m = 10$, $\Sigma = \{a, c, g, t\}$, $gap = 5$, this model shows the prediction of approximation ratio ε of SAIL algorithm is about 0.878. Therefore, this model can be used in pattern mining showed as below.

PMWL pattern mining evaluation mechanism

Input: Given T, Σ, m, gap, support sup

Output: pattern P

As we know, mining algorithm strategy is to learn from the strategy of matching algorithm, so PMWL pattern mining problem is naturally based on PMWL matching problem. For example, in mining algorithm MAIL (Xie et al., 2010), although a graph structure is utilized which conducts it different from SAIL; it is still based on the *left-most* strategy. As a result, they have the same degree of completeness. Therefore, our model can propose an evaluation mechanism for mining.

4.2. The impact of pattern *rep* on completeness

In next part, we will put forward another important concept, named *rep*, and analyze its impact on completeness. We first give an example to illustrate the reason why this concept is needed. Given $m = 4$, $\Sigma = \{a, c, g, t\}$, $gap = 2$, the corresponding patterns maybe $P_1 =$ a¢[0,2]c¢[0,2]g¢[0,2]t or $P_2 =$ a¢[0,2]c¢[0,2]c¢[0,2]t. They have the same Σ, m and gap. However, when applying SAIL or BPBM, the completeness of solutions is not the same, since for P_1 algorithms can obtain complete solutions while for P_2 can not.

Considering two examples below:

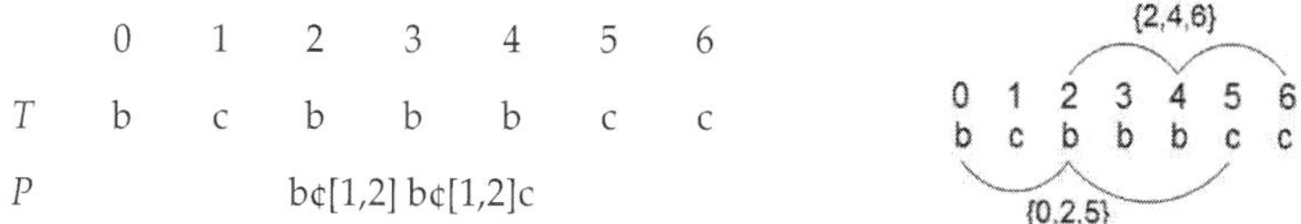

Table 9. Example 1 for *rep* concept

A complete occurrence set of this example is {{0, 3, 5}, {2, 4, 6}}, the number of matching occurrences is 2. It is not difficult to find that, in SAIL algorithm, for *position* 5, the selection of position 2 as $p[1]$'s occurrence by the *left-most* strategy will consume the position for the next matching occurrence. We can guess that, the recurring 'b' character in this pattern affect the quality of matching occurrences.

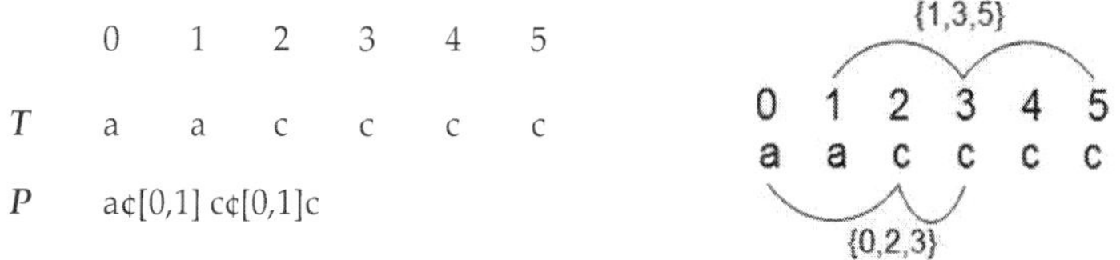

Table 10. Example 1 for *rep* concept

In this example, A complete occurrence set is {{0, 2, 4}, {1, 3, 5}}, the number of matching occurrences is 2. If we use SAIL algorithm and first obtain {0, 2, 3}, then we will only get this occurrence and lose {0, 2, 4}, {1, 3, 5}. Obviously, the recurring 'b' character in this pattern affects the completeness.

$P = p[0], p[1]… p[i]… p[m-1]$, $p[i]$ is stand for the *ith* letter in P where $i = 0,1,2……m-1$

$A = a[0], a[1]… a[u]… a[m-1]$, $a[i]$ is stand for the *ith* character maching position of occurrence A where $i = 0,1,2…m-1$

$B = b[0], b[1]… b[w]… b[m-1]$, another occurrece.

$S = s[0], s[1]… s[i]… s[k]… s[m-1]$, another occurrece.

Let $a[u], b[w]$ be the positions in A, B, which conflict with $s[i], s[k]$ in S *separately*. We assume other positions in A and B do not conflict with the occurrences in U_{SAIL}. $\therefore a[u] = s[i], b[w] = s[k]$ $\therefore t[a[u]] = t[s[i]], t[b[w]] = t[s[k]]$ $\because$ According to the definition 4, $t[a[u]] = p[u], t[s[i]] = p[i], t[\,b[w]\,] = p[w], t[s[k]] = p[k]$ $\therefore p[u] = p[i], p[w] = p[k]$

It would be discussed in the following two cases:

① $u \neq i$ or $w \neq k$

② $u = i$ and $w = k$

For ①, when if $u \neq i$, $\because p[u] = p[i]$ $\therefore$ There are two of the same letters from different positions in P. $\therefore$ According to definition 6, P is an R pattern. For the case of $w \neq k$, similarly, it can be proved.

Then we will prove the other condition is impossible, and conclude P is R pattern.

For ②, we obtain $a[u] = s[i] = s[u], b[w] = s[k] = s[w]$. There is $u \neq w$. $\because$ Assume $u = w$, then $u = i = w = k$ $\therefore a[u] = s[i] = s[k] = b[w]$. Consider A, B belong to the same occurrence set, which contradicts with LEMMA 1 $\therefore u \neq w$. Without loss of generality, let $u < w$, according to LEMMA 2 $\because$ SAIL adopts the *left-most* strategy, and $S \in U_{SAIL}, A, B \notin U_{SAIL}$ $\therefore s[u] < b[u], s[w] < a[w]$ $\because a[u] = s[u], b[w] = s[w]$

$$\therefore a[u] < b[u], b[w] < a[w] \tag{1}$$

And $\because u < w$, we can obtain $b[u] < b[w]$

$A = …a[u]……………..a[w]…$

$B = ……..b[u]…b[w]………$

The occurrence $\{b_0, b_1,…, b_u,…,a_w,…, a_{m-1}\}$ can be considerd as $\{\{b_0, b_1,…,b_u\}, \{b_u,…,a_w\}, \{a_w,…, a_{m-1}\}\}$.

According to the definition 4, $\{a_0, a_1,…,a_u,…,a_w,…,a_{m-1}\}$ and $\{b_0, b_1,…,b_u,…,b_w,…,b_{m-1}\}$ satisfy the local constraints. So $\{a_w,…, a_{m-1}\}$ and $\{b_0, b_1,…,b_u\}$ satisfy the local constraints.

Due to $\{b_u,…,a_w\}$, we can get $\{ b_u, b_{u+1}…, a_{w-1}, a_w \}$.

From the equation (1), $a[u] < b[u], b[w] < a[w]$, and according to the definition 4:

$$b[i] < b[i+1], a[i] < a[i+1] \text{ where } u \leq i \leq w-1 \tag{2}$$

There is a t satisfying:

$$b[t+1] < a[t+1] \text{ and } b[t] < b[t] \text{ where } u \le t \le w\text{-}1 \tag{3}$$

$$A = \ldots a[u]\ldots\ldots a[t]\ldots\ldots\ldots\ldots a[t+1]\ldots\ldots\ldots a[w]\ldots$$

$$B = \ldots\ldots\ldots b[u]\ldots\ldots b[t]\ldots\ldots b[t+1]\ldots\ldots b[w]\ldots\ldots\ldots$$

Assume there is no t satisfying the condition, consider $b[t] \ne a[t]$ where $u \le t \le w$. Then due to any t there is $a[t+1] < b[t+1]$ or $a[t] > b[t]$ where $u \le t \le w\text{-}1$. Consider $a[u] < b[u]$, there is $a[u+1] < b[u+1]$. Due to $a[u+k] < b[u+k]$, we can obtain $a[u+k+1] < b[u+k+1]$ where $0 \le k \le w\text{-}u\text{-}1$. Then we can induce $a[i] < b[i]$ where $u \le i \le w$. It contradicts $b[w] < a[w]$, so the assume is incorrect Due to equation (2) and (3), $a[t] < b[t] < b[t+1] < a[t+1]$ where $u \le t \le w\text{-}1$.

$$b[t+1] - b[t] < a[t+1] - b[t] < a[t+1] - a[t] \tag{4}$$

That is $a[t]$ and $b[t\text{-}1]$ satisfy the local constraints. In this way, $\{\ b_u,\ b_{u+1}\ldots,\ a_{w\text{-}1},\ a_w\ \}$ can be considered as $\{\{\ b_u,\ b_{u+1}\ldots,b_{t\text{-}1}\},\{b_t,\ a_{t+1}\},\{\ a_{t+2}\ \ldots,\ a_{w\text{-}1},\ a_w\ \}\}$. In accordance with definition 4, $\{\ b_u,\ b_{u+1}\ldots,b_{t\text{-}1}\},\{\ a_{t+2}\ \ldots,\ a_{w\text{-}1},\ a_w\ \}$ satisfy the local constraints. $\therefore \{\ b_u,\ b_{u+1}\ldots,\ a_{w\text{-}1},\ a_w\ \}$ satisfy the local constraints. $\therefore$ From the above analysis, $\{b_0,\ b_1,\ldots,b_u,\ldots,a_w,\ldots,\ a_{m\text{-}1}\}$ satisfy the local constraints.

However, according to the theorem, the other positions in A,B do not conflict with U_{SAIL} except for $a[u]$, $b[w]$. That is, $\{b_0,\ b_1,\ldots,b_u,,a_w,\ldots,a_{m\text{-}1}\}$ satisfies the *one-off* condition. $\therefore \{b_0,\ b_1,\ldots,b_u,\ a_w,\ldots,\ a_{m\text{-}1}\}$ is another occurrence, and does not conflict with any occurrences in U_{SAIL}. But U_{SAIL} does not include this occurrence. It contradicts with LEMMA 3. Thus, condition ② is impossible. And from the analysis of ①, under the condition of the theorem, P must be R pattern. The theorem 1 is proved.

THEOREM 2 Given a text T, a pattern P, if P is NR pattern, then SAIL is complete.

Proof: It is the inverse negation of THEOREM 1. Apparently, THEOREM 2 is true.

THEOREM 3 Given a text T, a pattern P, if P is R pattern, then SAIL is incomplete.

Proof: It can be concluded from the analysis and example in section 2.

THEOREM 4 If the pattern fulfills *gap* = 0, SAIL is complete.

Proof: If *gap* = 0, the wildcard is a constant. For example a¢[1,1]c¢[2,2]c is converted into a¢c¢c¢c. There won't be any conflict or exist seizing between occurrences. SAIL will perform complete.

Experiment design[1]: $\Sigma= 4$, $m = \{5,7,9\}$, *gap* = [0,3], *rep* = $\{0,1,2,3,4,6,7,10,11,15,\ 21,28,35\}$. In each set of experiments, 20 patterns are randomly generated; the final result is the average.

Analysis of experimental results: with increment of *rep*, the curve of approximation ratio gradually decreases, followed by a slight increase. The reason for decline is that *rep* lead to more nested occurrences, resulting in a greater degree of the possibility of losing occurrences; the reason for the increscent is that larger *rep* can cause more extreme pattern. For instance, when $\Sigma = 4$, $m = 7$, $rep = 21$, patterns like $P_1 =$

---

[1] When Σ and m are determined, *rep* can only be some certain values, because *rep* has correlation with Σ and m

a¢[0,3]a¢[0,3]a¢[0,3]a¢[0,3]a¢[0,3]a¢[0,3]a which is difficult to find a special text containing nested occurrences of such pattern, will be produced. For P_1, the text like "aaaaaaaaaaaaaa" contains nested occurrences of this pattern. Obviously, this extreme text is very rare. Therefore, under the premise of nested occurrences are not easily to be formed, the approximation ratio will be increased slightly.

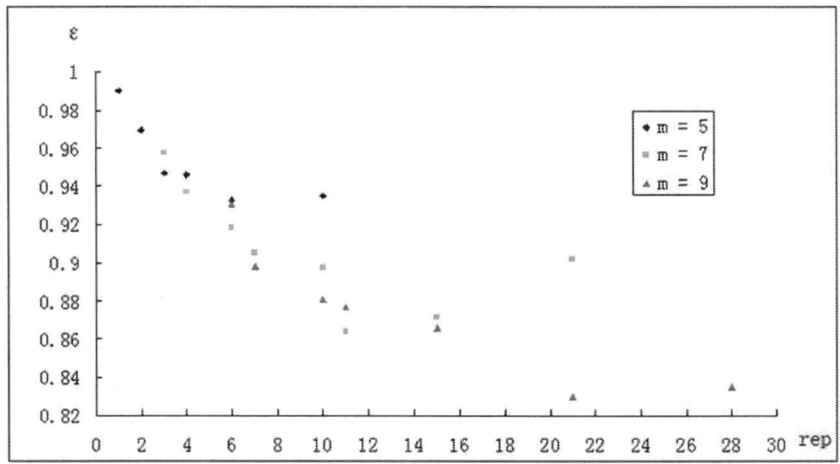

Figure 7. The relation between *rep* and approximation ratio ε

Next we will analyze the relationship between the repeatability *rep* and alphabet size Σ, pattern length m. Original problem: a pattern which length is m, and alphabet size is Σ, what is the expectation of repeatability E(*rep*)?

This description is equivalent to the model of 'taking ball from the bag' in the combination mathematics:

There is a bag of balls, and $|\Sigma|$ kinds of colors, taking m balls from the bag with replacement, then in fetched balls, how many pairs of the same color?

Σ \ m	3	4	5	6	......	m
3	3/3	6/3	10/3	15/3	......	$C_m^2 / 3$
4	3/4	6/4	10/4	15/4	......	$C_m^2 / 4$
5	3/5	6/5	10/5	15/5	......	$C_m^2 / 5$
6	3/6	6/6	10/6	15/6	......	$C_m^2 / 6$
......	......	......	......	......	......	......
Σ	$3/|\Sigma|$	$6/|\Sigma|$	$10/|\Sigma|$	$15/|\Sigma|$	......	$C_m^2/|\Sigma|$

Table 11. The relationship between Σ, m and *rep*

Finally, we can deduce:

$$E(rep) = \frac{C_m^2}{|\Sigma|} \tag{2}$$

5. Conclusions

As an extension of traditional matching problem, the PMWL problem has aroused more and more attention because of its unique flexibility and complexity. Based on problem definition and drawing on research idea in traditional matching problem, this article introduces SAIL, RSAIL, SBO and BPBM which are representative algorithms for PMWL in three important respects: the data structures, the matching strategies and the characteristics of pattern. The article also analyzes the pros and cons of the above algorithms from the point of quality of the solution and time complexity, and gives experimental matching results by using real DNA data. Among them, the SAIL algorithm is the first to propose the method of solving PMWL problem, it uses the sliding window structure and the representative *left-most* matching strategy. This paper finds that in short patterns, the approximation ratio of SAIL is higher than 0.9, while in longer patterns, the occurrences obtained by SAIL are of poor quality; the quality of occurrences obtained by SBO is best, but its time consumption has a non-linear relationship with the length of text; BPBM utilizes bit parallel technology to improve the efficiency of matching greatly, but also is impact by the machine word; for pattern with repeated letters in tail, RSAIL uses symmetry to improve the quality of occurrences under certain conditions, thus providing a solving idea to PMWL problem, but in longer patterns and wilder gaps, the efficiency is not obvious.

Afterwards, this article focus on relationship between approximation ratio ε and alphabet size Σ, pattern length m, wildcards span gap and repeatability rep. Firstly, this article proposes the model $\varepsilon = F(\Sigma, m, gap)$, describing the functional relationship between pattern characteristics and approximation ratio approximately; secondly, this article proves PMWL's completeness under the conditions of $rep = 0$; finally, the relationship between the pattern features are also analyzed andm in addition, relationship that $E(rep) = \frac{C_m^2}{|\Sigma|}$ is proposed.

In future work, the formal description of the PMWL problem will be considered, in order to explain the complexity of the problem better, thus helping algorithm design and analysis for problem complexity.

Author details

Haiping Wang, Taining Xiang and Xuegang Hu
Hefei University of Technology, China

6. References

Amir, A., Aumann, Y., Landau, G., Lewenstein, M. & Lewenstein, N. (2000). Pattern matching with swaps, *Journal of Algorithms*, 37(2): 247-266

Amir, A. & Navarro, G. (2009). Parameterized matching on non-linear structures, *Information processing letters*, 109(15): 864-867

Baeza-Yates, R. & Gonnet, G. (1992). A new approach to text searching, *Communications of the ACM*, 35(10): 74–82

Bille, P., Gørtz, I. L., Vildhøj, H. & Wind, D. (2010). String matching with variable length gaps, *Proceedings of 17th SPIRE*, pp. 385–394

Brudno, M., Steinkamp, R. & Morgenstern, B. (2004). The CHAOS/DIALIGN WWW server for multiple alignment of genomic sequences, *Nucleic Acids Research*, 32: 41–44

Califf, M. E. & Mooney, R. J. (2003). Bottom-up relational learning of pattern matching rules for information extraction, *Journal of Machine Learning Research*, 4(6): 177-210

Chen, G., Wu, X. D., Zhu, X.Q., Arslan, A. N. & He, Y. (2006). Efficient string matching with wildcards and length constraints, *Knowledge and Information Systems*, 10(4): 399–419

Cole, J.R., Chai, B., Marsh, T. L., Farris, R. J., Wang, Q., Kulam, S. A., Chandra, D. M., McGarrell, D. M., Schmidt, T. M., Garrity, G. M. & Tiedje, J. M. (2005). The ribosomal database project(RDP-11): Sequences and tools for high-throughput rRNA analysis, *Nucleic Acids Research*, 33(1): 294-296

Cole, R., Gottlieb, L. A. & Lewenstein, M. (2004). Dictionary matching and indexing with errors and don't cares, *Proceedings of the 36th ACM Symposium on the Theory of Computing*, ACM Press, New York, NY, USA, pp. 91–100

Fischer, M. J. & Paterson, M. S. (1974). String matching and other products, *In Karp RM(ed) Complexity of computation, Massachusetts Institute of Technology*, Cambridge, MA, USA, vol 7, pp. 113-125

Gusfield, D. (1997). Algorithms on strings, trees and sequences: computer science and computational biology, chapter 6, Cambridge University Press

Guo, D., Hong, X. L., Hu, X. G., Gao, J., Liu, Y. L., Wu, G. Q. & Wu, X. D. (2011). A Bit-Parallel Algorithm for Sequential Pattern Matching with Wildcards, *Cybernetics and Systems*, 42(6): 382-401

He, D., Wu, X. D. & Zhu, X. Q. (2007). SAIL-APPROX: An efficient on-line algorithm for approximate pattern matching with wildcards and length constraints, *IEEE International Conference on Bioinformatics and Biomedicine (BIBM'07)*, IEEE Computer Society, pp. 151–158

He, Y., Wu, X. D., Zhu, X. Q. & Arslan, A. N. (2007). Mining Frequent Patterns with Wildcards from Biological Sequences [C], *IEEE International Conference on Information Reuse and Integration*, Las Vegas, IL, pp. 329-334

Ji, X. N., Bailey, J. & Dong, G. Z. (2007). Mining minimal distinguishing subsequence patterns with gap constraints, *Knowledge and Information Systems*, 11(3): 259-286

Kucherov, G. & Rusinowitch, M. (1995). Matching a set of strings with variable length don't cares, *Proceedings of the 6th Symposium on Combinatorial Pattern Matching*, Springer, Berlin Heidelberg New York, pp. 230–247

Manber, U. & Baeza-Yates, R. (1991). An algorithm for string matching with a sequence of don't cares, *Information Processing Letters*, 37(3): 133–136

Min, F., Wu, X. D. & Lu, Z. Y. (2009). Pattern matching with independent wildcard gaps, *Eighth IEEE International Conference on Dependable, Autonomic and Secure Computing(DASC-2009)*, Chengdu, China, pp. 194-199

Morgante, M., Policriti, A., Vitacolonna, N. & Zuccolo, A. (2004). Structured motifs search, *Proceedings of the 8th annual international conference on Computational molecular biology*, In print

Muth, R. & Manber, U. (1996). Approximate multiple string search, *Combinatorial Pattern Matching, Springer*, pp. 75–86

Muthukrishnan, S. & Krishna, P. (1994). Non-standard stringology: algorithms and complexity [C], *Proceedings of the twenty-sixth annual ACM symposium on Theory of computing New York*, NY, USA, pp. 770-779

"National center for biotechnology information website", [online], available: http://www.ncbi.nlm.nih.gov/

Navarro, G. & Raffinot, M. (2001). Flexible pattern matching in strings: practical on-line search algorithms for texts and biological sequences, Cambridge University Press

Rahman, M. S., Iliopoulos, C., Lee, I., Mohamed, M. & Smyth, W. F. (2006). Finding Patterns with Variable Length Gaps or Don't Cares, *Computing and Combinatorics, 12th Annual International Conference*, COCOON 2006, Taipei, Taiwan, August 15-18, Proceedings. Vol. 4112

Sagot, M. F. & Viari, A. (1996). A Double Combinatorial Approach to Discovering Patterns in Biological Sequence, *Proceedings of the 7th Symposium on Combinatorial Pattern Matching*, Springer, pp. 186-208

Wang, H. P., Xie, F., Hu, X. G., Li, P. P. & Wu, X. D. (2010). Pattern Matching with Flexible Wildcards and Recurring Characters, *Proceedings of 2010 IEEE International Conference on Granular Computing*, pp. 782-786

Wu, Y. X., Wu, X. D., Jiang, H. & Min, F. (2011). A Heuristic Algorithm for MPMGOOC, *Chinese Journal of Computers*, 34(8): 1452-1462

Xie, F., Wu, X. D., Hu, X. G., Gao, J., Guo, D., Fei, Y. L. & Ertian, H. (2010). Sequential Pattern Mining with Wildcards [C], *22nd IEEE International Conference on Tools with Artificial Intelligence (ICTAI)*, pp. 241-247

Zhang, M. H., Kao, B., Cheung, D. W. & Yip, K. Y. (2005). Mining periodic patterns with gap requirement from sequences, *Proceedings of ACM SIGMOD*, Baltimore Maryland, pp. 623–633